SPATIAL
MATHEMATICS

Theory and Practice
through Mapping

SPATIAL MATHEMATICS

Theory and Practice through Mapping

Sandra Lach Arlinghaus
Joseph J. Kerski

CRC Press
Taylor & Francis Group
Boca Raton London New York

CRC Press is an imprint of the
Taylor & Francis Group, an **informa** business

CRC Press
Taylor & Francis Group
6000 Broken Sound Parkway NW, Suite 300
Boca Raton, FL 33487-2742

First issued in paperback 2019

ISBN-13: 978-1-4665-0532-2 (hbk)
ISBN-13: 978-0-367-86704-1 (pbk)

Library of Congress Cataloging-in-Publication Data

Arlinghaus, Sandra L. (Sandra Lach), author.
    Spatial mathematics : theory and practice through mapping / Sandra Lach Arlinghaus, Joseph J. Kerski.
        pages cm
        Summary: "Spatial mathematics and analysis, two different approaches to scholarship, yield different results and employ different tools. This book explores both approaches to looking at real world issues that have mathematics as a critical, but often unseen, component. Readers learn the mathematics required to consider the broad problem at hand, rather than learning mathematics according to the determination of a (perhaps) artificial curriculum. This format motivates readers to explore diverse realms in the worlds for geography and mathematics and in their interfaces"-- Provided by publisher.
        Includes bibliographical references and index.
        ISBN 978-1-4665-0532-2 (hardback)
        1. Geography--Mathematics. 2. Cartography--Mathematics. 3. Geographic information systems--Software. I. Kerski, Joseph J., author. II. Title.

G70.23.A77 2013
910.01'51--dc23   2012049508

Visit the Taylor & Francis Web site at
http://www.taylorandfrancis.com

and the CRC Press Web site at
http://www.crcpress.com

Contents

Update Bank: http://www-personal.umich.edu/~sarhaus/
SpatialMathematics/index.html

Preface

A book on "spatial mathematics" is a natural first-choice in this series from CRC Press entitled Cartography, GIS, and Spatial Science: Theory and Practice. Mathematics is the theoretical backbone of computer science and therefore of its derivatives such as mapping software. How nice it is to have the opportunity to turn things around...to use mapping to help motivate readers to follow some of the intricacies of mathematics and to stimulate them to look for interactions between mathematics and geography (and elsewhere) according to their interests.

P.1 Why spatial mathematics?

As the world becomes increasingly measured and analyzed, more and more of those measurements and analyses are being placed into a compact, understandable medium—the map. Today's maps are not static—they are dynamic, changing with traffic, weather, and wildfires. They are no longer the product of long-established mapping agencies: Ordinary citizens can make changes to them. These maps are instantly accessed on our smartphones, in our vehicles, and on our computers. Everyday phenomena and objects are tracked and mapped, from containers on ships to city buses to thunderstorms—even your smartphone itself is mapped. These maps are not mere graphics—they are tied to powerful and interconnected databases in a Geographic Information Systems (GIS) environment. It is more important than ever to understand the mathematics behind the rapidly expanding array of maps and data. Furthermore, the mathematics we use in the real world often influences the decisions we make. For example, it influences how we georeference mapped data because it influences how the data sets were collected and processed in the first place. The topic of "mapematics and decisions" is a valuable educational tool and a critical one for understanding spatial elements of real-world process—from politics, to planning, to economics. Indeed, a recent study from the University of Chicago shows that spatial knowledge enhances understanding of numbers at even early ages (*Science Daily*, 2012; Toddler Spatial Knowledge Boosts Understanding of Numbers).

In 1989, the National Council of Teachers of Mathematics (NCTM) released its *Curriculum and Evaluation Standards for School Mathematics*. This document provided guidance for developing and implementing a vision of mathematics and instruction that serves all students. In 2000, NCTM expanded and elaborated on the 1989 standards to create *Principles and Standards for School Mathematics*. The discussions in this book adhere to numerous NCTM content standards, including Number and Operations, Algebra, Geometry, Measurement, and Data Analysis. The activities in this book connect to several NCTM Process Standards, including problem solving, reasoning and proof, communication, connections to other disciplines (in particular, geography, economics, and Earth science), and representation. The use of GIS supports all of the key standards featured as NCTM illuminations (http://illuminations. nctm.org/Standards.aspx). These include numbers and operations, algebra, geometry, measurement, data analysis and probability. Specifically, these include working with vectors (direction and length), perimeter and area; working with cell-based raster data; and working on estimation, units of measure, shapes and patterns, speed and distance, modeling, percentages, and ratios.

P.2 Why mapping?

In the chapters in this book, the content and skills presented are connected to other areas of mathematics, to other disciplines, and to real-world applications. Today's mapping is done in a GIS environment. GIS represents a systems approach and a holistic approach to examining the world. People using GIS organize, record, and communicate information and ideas using words, maps, tables, and symbols, interpreting information presented in various forms.

We believe that GIS provides an excellent way to teach mathematical concepts and skills through the visualization of numbers. Representing numbers, understanding patterns, relationships, and function, two-dimensional and three-dimensional geometric and spatial relationships, probability, statistics, change, models, measurements, problem solving, reasoning, connections, and communications are critical concepts. Every one of these can be explored using GIS tools and methods. Comparing graphs and maps of birth and death rates over time and region, analyzing the response of a stream to a recent storm through a real-time hydrograph, and creating cross-sections of terrain are three common activities in geography instruction, easily done in a GIS environment (**Figure P.1**).

All of them—and thousands more geographic activities—involve analyzing numbers. One might say that the core of GIS work is actually visualizing numbers, since the basis for mapping is to represent numbers as cells, points, lines, or polygons on a map. For example, the map in **Figure P.1** represents the results of the analysis of rainfall seasons and amounts in South Asia using ArcGIS software. When we connect latitude and longitude to the Cartesian coordinate system, when we measure area, shape, size, and distance in different map projections, when we compare geometric to exponential growth rates of agricultural output, even when we explain the Earth's shape, rotation, and revolution, we are applying geographic and mathematical concepts and can use GIS to teach it.

First, the use of GIS brings math to life by making math visual. Think of common problems such as: "Where and when will these trains cross paths? One departs Point *A* at 6:00 a.m. and heads toward Point *B* at 70 kilometers per hour, while the other departs Point *B* at 6:30 a.m. bound for Point *A* at 60 kilometers per hour." Where and when will they cross paths?" GIS allows the anchoring of these and other problems in the real world: Points *A* and *B* could be Cheyenne and Casper, and students can determine and plot the course and passing point at a real point on a GIS map layer.

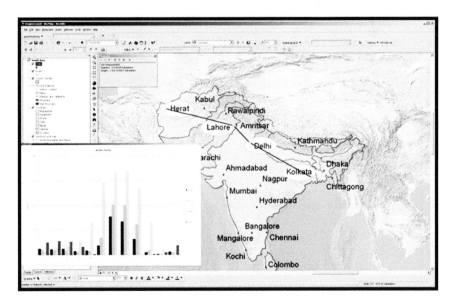

Figure P.1 *Mapped data set displayed in GIS environment. Source of base map: Esri software.*

Second, solving math problems in a GIS environment allows students to grapple with biodiversity, crime, natural hazards, climate, energy, water, and other relevant real-world issues of the twenty-first century. For example, the map below (**Figure P.2**) shows the analysis of the elevation of oil and gas wells in western Colorado.

Third, students often do not feel that what they are learning is relevant to what they will be doing after they graduate. Hundreds of jobs in geospatial technologies—not just surveyors and remote sensing analysts—require analytical, statistical, and computational skills that are learned in mathematics.

The National Governors Association Center for Best Practices and the Council of Chief State School Officers in 2009 began releasing common core standards for public comment. Beginning with career and workforce-readiness standards, it continued in 2010 with the release of English Language Arts and Mathematics standards. As these standards are aligned with college and work expectations, apply higher-order skills, and especially because they all revolve around problem-solving, we believe that a strong case can be made through the common core initiative why teaching with the spatial perspective is an essential part of primary, secondary, higher, and informal education. Furthermore, the math core content standards include such items as

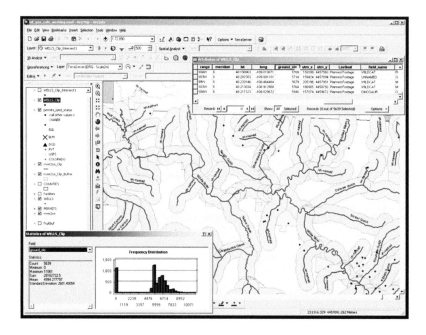

Figure P.2 *Elevation of oil and gas wells in western Colorado. Source of base map: Esri software.*

attending to precision, and constructing viable arguments, both of which are core to GIS.

Mathematically proficient students understand and use stated assumptions, definitions, and previously established results in constructing arguments. They make conjectures and build a logical progression of statements to explore the truth of their conjectures. They break things down into cases and can recognize and use counterexamples. They use logic to justify their conclusions, communicate them to others, and respond to the arguments of others. They reason inductively about data, making plausible arguments that take into account the context from which the data arose. They make sense of complex problems and persevere in solving them.

Mathematically proficient students start by explaining to themselves the meaning of a problem and looking for entry points to its solution. They consider analogous problems, try special cases, and work on simpler forms. They evaluate their progress and change course if necessary. They try putting algebraic expressions into different forms or try changing the viewing window on their calculator to get the information they need. They look for correspondences between equations, verbal descriptions, tables, and graphs. They draw diagrams of relationships, graph data, search for regularity and trends, and construct mathematical models. They check their answers to problems using a different method, and they continually ask themselves, "Does this make sense?" They look for and make use of structure. Mathematically proficient students look closely to discern a pattern, and GIS is all about investigating patterns. They make strategic decisions about the use of technological tools. GIS is a technological tool but one that encourages critical thinking about data and processes.

In terms of statistics, GIS offers many connections. With GIS, data are gathered, displayed, summarized, examined and interpreted to discover patterns. Data can be summarized by statistics involving measuring center, such as mean and median, and statistics measuring spread, such as interquartile range and standard deviation. Different distributions can be compared numerically and visually, using different classification methods, histograms, and standard deviational ellipses. Which statistical measures to use, the results of these considerations, depend on the problem investigated, the data gathered and considered, and the real-life actions to be taken.

If you are in a GIS-related career or are contemplating one, improving your mathematical skills will improve your job skills. The US Department of Labor Employment and Training Administration (ETA) worked with industry and education leaders to develop a comprehensive competency model for Geospatial Technology. Known as the Geospatial Technology Competency Model (GTCM) it specifies personal effectiveness, academic, workplace, industry-wide, industry-sector, management, and occupation-specific competencies that are important for success in the field (http://www.careeronestop.org/competencymodel/pyramid.aspx?geo = Y).

One of the eight key academic competencies is a firm grounding in mathematics. In addition, the ability to use such core GIS tools as geo-processing and spatial statistics hinge upon a solid understanding of the principles and applications of mathematics.

We, as authors, hope that all readers will find something here to expand their horizons and lead them further along their journey to proficiency in spatial mathematics! Enjoy reading the book as much as we have enjoyed writing it!

Sandra L. Arlinghaus
Ann Arbor, Michigan

Joseph J. Kerski
Broomfield, Colorado

Acknowledgments

The authors thank so many they have met who have shown an interest in the intersections of mathematics and geography for many years, co-authors of materials used to form a base for this book (cited in QR codes at the ends of chapters), co-authors of related materials, creators of software used in support of the ideas and ideals, colleagues in their respective workplaces, staff at the publishing company, and of course their families.

In particular, Arlinghaus wishes to thank collaborators who have shared co-authored materials: William C. Arlinghaus (Ph.D. Algebraic Graph Theory; Professor Emeritus Mathematics and Computer Science, Lawrence Technological University), William E. Arlinghaus (Cemeterian, President Greenscape Inc.), David E. Arlinghaus (student, Washtenaw Community College), and Diana Sammataro (USDA-ARS Carl Hayden Honey Bee Research Center). From Community Systems Foundation she thanks co-authors: John D. Nystuen (Ph.D. Geography; Professor Emeritus of Geography and Urban Planning, The University of Michigan; CEO of CSF) and Roger Rayle (Research Associate). For ongoing support, she thanks Kris S. Oswalt (President of CSF) and Sandra S. Westrin (Senior Administrator). From The University of Michigan, School of Natural Resources and Environment thanks go to: Professor Paul Mohai (Ph.D.), Professor and Former Dean Rosina Bierbaum (Ph.D.), and Professor and Dean Marie Lynn Miranda (Ph.D.). For inspiration of various sorts, thanks go to Matthew Naud, Wendy Rampson, Adele El-Ayoubi (City of Ann Arbor); Janis Bobrin and Harry Sheehan (Washtenaw County) as well as to Karen Hart, Alma S. Lach, Ann E. Larimore, and Gwen L. Nystuen.

Kerski wishes to thank all those who teach and foster spatial thinking in education and society through the use of GIS and other geospatial technologies. He believes they are making a positive difference on our planet and on people's lives. He also wishes to thank Esri for the innovative tools they create.

Both authors thank those who have kindly granted permission for use of materials (noted in captions of figures). Both authors extend greatest thanks and appreciation to the outstanding staff at CRC Press/Taylor & Francis Group, L.L.C. In particular, we worked most closely with Irma Shagla Britton, Editor for Environmental Sciences and Engineering, who stayed with us throughout the process and offered wise counsel and friendly advice. Her support and encouragement were outstanding! Edward Curtis, Project Editor, provided excellent support in helping us get our original document into final form; his caring and careful hand demonstrated his fine art in document creation and in communication about document creation. Other helpful individuals we met along the way included: Randy Burling, Manager, Production Editing; Joselyn Banks-Kyle, Project Coordinator, Editorial Project Development; Laurie Schlags, Project Coordinator; Arlene Kopeloff, Editorial Assistant; and, Nora Konopka, Publisher of Engineering and Environmental Science, all at Taylor & Francis. They have been forward-looking, enthusiastic, and particularly supportive of the imaginative aspects of this project. We send them our greatest thanks.

We alone are responsible for any blunders, errors, or unintended consequences that remain in the document. We do wish to thank everyone who has contributed directly or indirectly to this effort and while we issue great thanks to many we do hope that we have not missed anyone!

- *Software used:* Adobe Photoshop; ArcGIS, versions 10.1 and earlier; Java Development Kit; Microsoft Office: Word and Excel; Microsoft Windows 7 and Math Input Panel; QR code generator, http://createqrcode.appspot.com/; Word cloud generator, http://www.wordle.net/
- *Hardware*: computers running Windows XP and Windows 7.
- *Cover art*: Eric Gaba, June 2008. Data: US NGDC World Coast Line (public domain). Wikipedia reference: http://en.wikipedia.org/wiki/File:Tissot_indicatrix_world_map_sinusoidal_proj.svg
- *Title page art*: Public domain image produced by Paul B. Anderson, available at: http://www.geowebguru.com/articles/202-choosing-a-projection-part-2-pseudo-conic-azimuthal-and-cylindrical-projections/

Introduction

Imagination is more important than knowledge.

Albert Einstein

Everything is related to everything else, but near things are more related than distant things.

Waldo Tobler
First Law of Geography

In today's world, maps are numbers first, pictures later...and then we build on the numeric aspects of GIS with math bridges.

Joseph Berry

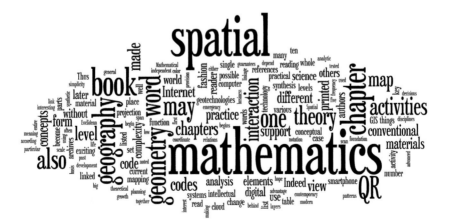

1.1 Scope of the work

What is "spatial mathematics"? Spatial mathematics draws on the theoretical underpinnings of both mathematics and geography. Spatial mathematics draws not only from all forms of geometry, but also from topology, graph theory, trigonometry, modern algebra, symbolic logic, set theory, and others. The material in this single book simply cannot do justice to the broad arena in which this interaction takes place. We draw on concepts from various subfields but in no way attempt comprehensive coverage—well beyond the scope of this small single volume (later volumes may focus on other subfields). Rather, the material in this book is designed to indicate how to build the bridges between mathematics and geography. To do so, it is critical to begin each chapter with theoretical discussions that form the bridge foundation, and activities that form the span between the two disciplines. As is often the case in forging such interaction, the level of the theory is more advanced than is the level of the practice; interpretation of theory in the real-world must be done from a vantage point that is deeper in perspective than the application itself.

To communicate some of the key concepts and illustrations, we use current technology (bearing in mind that its currency is of course a moving target) and employ QR codes (an extra dimensional bar code) to span the digital divide between the conventionally-printed word and the capability of the Internet to provide additional meaning. That meaning may stem from an animation or visualization that printed pages cannot support or another piece of information that provides more detail or a different way of looking at things. What had once been "distant things," in the manner of Tobler (1970), may now become "near things." The Internet, smartphones, and other devices offer an interesting challenge to interpreting Tobler's First Law of Geography.

Take your smartphone, enabled with a QR code reader, and scan the pattern to link to an animated map, to a mapping activity site, to an interactive webpage, or a host of other possibilities—all without a laptop or other direct computer access. Imagine sitting in the airport and preparing for class by working on Internet mapping activities using your smartphone and this book. E-mail an interesting link to yourself for later detailed consideration on your computer (or if you come across a link not supported by your smartphone). The use of this particular form of contemporary technology makes this book uniquely suited for self-study. Indeed, the cover of the book itself has a QR code on it that will come to life as the authors tell you, in a brief video, what they view to be the significance of this sort of work!

The interaction of spatial science and mathematics is nothing new. Eratosthenes of Alexandria was not only the Librarian of the Great Library of antiquity but also made leading discoveries in number theory, with his prime number sieve and in spatial science with his amazingly accurate measurement of the circumference of the Earth, long before space travel was even dreamed of! Later, names such as Copernicus (1543), Kepler (1609), Varenius (1650), Leibniz (1686), and Newton (1687) are associated with the linkage between mathematics and spatial science. Into the twentieth century, among others, Albert Einstein (physics—1916), D'Arcy Thompson (zoology—1917), Kurt Lewin (psychology—1936), and Claude Levi-Strauss (anthropology—1948) noted similar interactions between mathematics and conceptual elements of their own disciplines. Willis Ernest Johnson wrote *Mathematical Geography* back in 1907. In the 1960s, members of the Michigan Inter-University Community of Mathematical Geographers drew on such interaction on a regular basis. Indeed, William Warntz (1965) notes, "Bunge has pointed out the necessity and the efficiency of recognizing the inseparability of geometry as the mathematics, i.e., the language, of spatial relations and geography as the science of spatial relations." As Mac Lane later noted, it is the "fit of ideas" that is critical (1982).

Such inseparability also appears not only in the association of geography and geometry but also in the balance between "analysis" and "synthesis." Today's fashion is to talk about spatial analysis, and most of the computer programs that involve interaction between mathematics and mapping are based on analytic geometry and coordinate systems. Broadly viewed, "synthesis" assembles a whole from its parts while "analysis" begins with the whole and zooms in to the parts. Classical Euclidean geometry is one form of synthetic geometry, and, to be sure, it underlies the foundation of analytic geometry. However, very little contemporary use is made of the synthetic geometry itself, either in Euclidean or non-Euclidean form (Coxeter, 1961a and 1961b). Even less is made of "spatial synthesis." The last chapter of the book hints at exciting future directions in the context of past and present views of the interaction between geometry and geography.

The birth of Geographic Information Systems (GIS) technology during the "quantitative revolution" in geography during the 1960s and its growth in natural resource management, business, and planning in government, business, nonprofit organizations, and academia through the first decade of the twenty-first century was in large part due to the development of models and algorithms in geography and in mathematics. GIS, together with GPS and remote sensing, collectively known as the geotechnologies, comprise one of the three fastest career growth fields according to the US Department of Labor. Today, geotechnologies have expanded from the professional community to apps and tools that ordinary citizens can take advantage of in their vehicles, on their smartphones, and on their tablets and laptops. This expansion was made possible by the development and deployment of configurable Application Program Interfaces (APIs) powered by JavaScript, HTML, Python, and other scripting languages, and again it relies heavily on mathematics to display, query, and modify spatial information through digital maps. Indeed, it could be said that modern society has been made possible through the intersection of geography and mathematics, since virtually all airline flights, rapid transit, shipping and trucking, emergency vehicle response, agriculture, city planning, wastewater treatment, provision of electricity, and many other infrastructures and systems depend upon the geotechnologies.

Indeed, just as there would be no GIS without the mathematics behind geoprocessing tools and of the very storing of spatial data in vector and raster formats; there would be no spatial analysis in geography without mathematics. Likewise, mathematics would be vastly inhibited without considerations of scale, ratios, angles, directions, patterns, projections, and other spatial concepts.

1.2 Structure of the book

When reading mathematical documents (and perhaps others as well), it is helpful to look over the entire document (book, in this case) and get a sense of the big picture. If one starts by reading at a detailed level, it is easy to become lost. Read at different levels. Start with the general and then move to the specific. That is, read in a hierarchical fashion first, and only after that then read in a linear fashion within layers (book chapters).

Each of the ten chapters has a section on "theory" and a section on "practice." We have attempted to write the material in a fashion that is intuitive rather than laden with a lot of notation. Often, the underlying notation is no more than a click away. In **Chapter 9**, for example, scan the QR code to see the entire USGS Map Projection poster. This book is written in layers, in support of the view of reading at both the general and specific level. The reader is encouraged to think about the deep connections that often underlie what we intend to be simply stated language. Do not mistake simplicity for simple-mindedness—we view simplicity as the hallmark of elegance. Thus, we encourage readers not only to appreciate such a viewpoint but also to embrace it and adopt it within their own intellectual portfolio.

The **Table of Contents**, listing headings and subheadings of each chapter, serves to describe the book. In the early part of the book, refresh your understanding of the geometry of the sphere and support it with a quick, spatial view of trigonometric functions. Examine coordinate systems, of various sorts, used to capture different elements of the Earth's surface. Move on to consider the role of mathematical transformations in a variety of real-world contexts. Expand your knowledge of color. How many colors suffice to color a map in the plane? How many on the sphere? Think about the algebra of pixels. Extract principles for making one type of map (dot density map) from a systematic view of map scale. Consider that choices made in the partitioning of data sets can influence subsequent decisions. Visualize spatial hierarchies. Think about analyzing distributional patterns behind emergency services provision. Ensure that you are grounded in making good decisions regarding some basic elements of map projection. Finish by thinking not only about past and current spatial interaction between mathematics and geography but also about exciting glimpses of possible futures. In all cases, support concepts with tested practice activities. Notice that one activity, "Siting an Internet café in Denver," is woven across several chapters. While each chapter does contain independent activities in support of the concepts of that particular chapter, the carrying of one activity as a persistent theme reflects not only the interdisciplinary character of spatial mathematics itself but also the simple fact of life that answers obtained in one situation may well play into outcomes in other situations.

Each of the ten chapters begins with a word cloud, as an abstract summary of the content of the chapter. It embodies a conventional journal "abstract," a conventional list of "keywords," and a conventional word frequency count thereby unifying them in a single image. Within the word cloud, the size of the word represents the number of occurrences of the word in the chapter. Beneath each of the chapter word clouds there is a list of five keywords selected from the word frequency counts associated with the word cloud for that chapter—the five most frequently used technical words are chosen. The word set in the glossary (intended for the non-expert), however, is selected by hand by the authors from the entire set of 10 chapters; the whole is greater than the sum of its parts.

Each of the 10 chapters ends with a table containing QR codes and hyperlinks leading to more conceptual and practical materials, drawn primarily from the substantial, independently developed, online archives of each author. Many of the QR codes and hyperlinks listed here have served as one basis for the chapter after which they are listed. This table of QR codes and hyperlinks is followed by a list including other inline references and additional related readings. Because the archives are complex, it should be an advantage to the reader to have the authors who created the archives guide and select, via QR codes and hyperlinks, conceptual references and practical activities from the linked sets. The practical activities are rated according to complexity, with 1 low and 5 high. Generally, higher levels of complexity require greater background or intellectual maturity than do lower levels; **Chapters 1** and 2 are in some ways "easier" than later ones, in an effort to ensure that we are all together, at least at the beginning. Nonetheless, even beginners should be able to derive benefit from some of the more advanced activities. We hope all the activities pique interest, independent of the level of complexity. Thus, it is not problematic to begin at a low level of complexity while building confidence.

As the printed materials preceding the table suggest one linkage between theory and practice (as hinted at in **Figure P.3**), the materials in the QR code and hyperlink tables of theory and practice greatly enlarge the tip of the iceberg presented in the printed chapter. Please use them and encourage others using the book to read it with smartphone in hand! Many of the linked materials are housed on servers with some degree of persistence—either through corporate or educational websites. All links from urls and from QR codes were tested and functional as of January 2013. Conventional references in a traditional bibliography may go out of print and become unavailable; interlibrary loan can be helpful in that regard. So, too, linked materials may change. Links that function today may not function tomorrow. Java Applets may function differently on different platforms and in different browsers. They also depend on various browser settings. Thus, we include screen shots. The world of the Internet is a rich and exciting place. It is not, however, a perfect place. A big difference between the conventional and digital issues is speed of change; as is the case in other regards, the digital world may change more swiftly than the conventional printed world. Therefore, the reader is encouraged to stay current, not only with changes in URLs, but also in the world of spatial mathematics, through a living web resource provided by the authors, as listed in the **Table of Contents**.

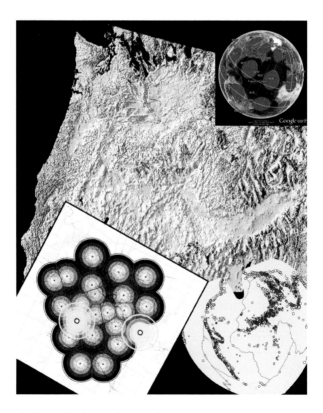

Figure P.3 *Collage of selected figures from this book.*

So there are no guarantees of permanence in electronic format; but then again there are no guarantees of permanence with a printed document. The only thing we hope we can guarantee is that you will find excitement, interest, and intellectual stimulation when you join with us on a voyage through some of the elements of spatial mathematics!

I.3 Related theory and practice: Access through QR codes

Theory

Persistent archive:

University of Michigan Library Deep Blue: http://deepblue.lib.umich.edu/handle/2027.42/58219

From Institute of Mathematical Geography site: http://www.imagenet.org/

Arlinghaus, S. 2012. Visual Abstracts: Institute of Mathematical Geography. Solstice: An Electronic Journal of Geography and Mathematics. Volume XXIII, Number 1. http://www.imagenet.org/

Practice

From Esri site: http://edcommunity.esri.com/arclessons/arclessons.cfm

Kerski, J. Connecting GIS to STEM (Science Technology Engineering and Mathematics) Education. http://edcommunity.esri.com/arclessons/lesson.cfm?id=695. Complexity level: 3.

Kerski, J. Connecting GIS to Environmental Education. http://edcommunity.esri.com/arclessons/lesson.cfm?id=693. Complexity level: 3, 4, 5.

Kerski, J. Why GIS In Education Matters. http://edcommunity.esri.com/arclessons/lesson.cfm?id=696. Complexity level: 1, 2, 3, 4, 5.

Archives

Sandra L. Arlinghaus, Ph.D., sarhaus@umich.edu
Adjunct Professor of Mathematical Geography and Population-Environment
 Dynamics, School of Natural Resources and Environment, The University of
 Michigan, Ann Arbor, MI, 48109 USA.
Personal Home Page: http://www-personal.umich.edu/~sarhaus/

Joseph J. Kerski, Ph.D., jkerski@esri.com
Education Manager
Esri
1 International Court, Broomfield CO 80021-3200, USA.
Personal Home Page: http://www.josephkerski.com

Institute of Mathematical Geography,
Deep Blue Archive Site.
Persistent online archive of
The University of Michigan.
http://deepblue.lib.umich.edu/handle/2027.42/58219

Community Systems Foundation, Home Page.
http://www.csfnet.org

Authors

Sandra L. Arlinghaus is a mathematical geographer by training who believes that pure mathematics, particularly geometry, is the key to the understanding and the creation of both spatial analysis and spatial synthesis. She holds a Ph.D. in theoretical geography (The University of Michigan) and other degrees and advanced education in mathematics (from Vassar College, the University of Chicago, the University of Toronto, and Wayne State University). She has published over 300 books and articles, many in peer-reviewed publications. She continues other innovative approaches in publication as creator of *Solstice: An Electronic Journal of Geography and Mathematics*, perhaps the world's first online peer-reviewed publication (1990-), and as co-creator with William C. Arlinghaus and Frank Harary of John Wiley & Sons' first e-book in 2002. Her interests are wide-ranging and interdisciplinary as is her teaching experience: the latter including innovative spatial approaches to remedial mathematics (The Ohio State University), to serving as a professor of mathematics teaching a wide range of beginning and advanced undergraduate courses, to creating spatially-oriented environmental science courses at both the beginning and advanced levels, to developing spatial projects in a virtual reality engineering course. Currently, she is Adjunct Professor of Mathematical Geography and Population-Environment Dynamics in the School of Natural Resources and Environment at The University of Michigan. In addition, she is a member of the Board of Trustees and Executive Committee of Community Systems Foundation, an international non-governmental organization (NGO). She also enjoys history, zoology, foreign languages, classical music, cooking, the planning of tournament duplicate bridge events, and community service, having served nine years on the City of Ann Arbor Planning Commission, five years on the City of Ann Arbor Environmental Commission, and five years on the Ann Arbor Police Department's Neighborhood Watch Citizen Advisory Panel.

Joseph J. Kerski is a geographer by training who believes that spatial analysis with GIS technology can transform education and society through better decision-making using the geographic perspective. He holds three degrees in geography, with emphases on computer cartography, GIS, population geography, and geographic education. He served for 22 years as geographer and cartographer at three US Federal Agencies, including the National Oceanic and Atmospheric Administration and Geophysical Data Center (NOAA), the US Census Bureau, and the US Geological Survey. During 2011, Kerski was President of the National Council for Geographic Education after serving on the board of directors for the organization for 10 years. He teaches online and face-to-face GIS and geography courses at primary and secondary schools, in community colleges, and in universities. He currently holds a position as Adjunct Instructor at the University of Denver. He currently serves as Education Manager for Environmental Systems Research Institute (Esri), focusing on GIS-based curriculum development, research in the implementation and effectiveness of GIS in education, teaching professional development institutes for educators, and fostering partnerships and communication to promote and support spatial analysis in formal and informal education at all levels, internationally.

Geometry of the Sphere

Keywords: Earth (46), GPS (40), longitude (38), latitude (38), sphere (36)

> The brain is wider than the sky,
> For put them side by side;
> The one the other will include,
> With ease, and You beside.
>
> **Emily Dickinson**

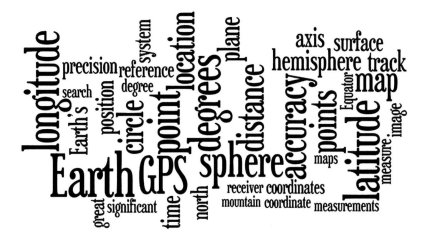

1.1 Introduction

A logical starting point for investigating the surface of the Earth is to examine its general shape—an approximate sphere. The sphere looks like a simple object, having no holes, no twists, and a generally smooth and uniform appearance. The geometry of the Earth as a sphere is anything but simple.

1

Moreover, the Earth is not exactly a sphere. When land masses, weather patterns, ocean currents, and other terrestrial phenomena are overlain on the close-to-spherical Earth, the underlying geometry becomes even more complex. Often it is helpful to look at the geometry, and the Earth itself, from the synthetic as well as the analytic viewpoint.

The whole is often more than the sum of the considered individual parts. Analysis begins with a global view and dissects that view to consider individual components, as they contribute to that whole. In contrast, synthesis often begins with a sequence of local views and assembles a global view from these, as a sum of parts. The assembled view may or may not fit "reality." These two different approaches to scholarship yield different results and employ different tools. In this book, we consider both approaches to looking at real world issues that have mathematics as a critical, but often unseen, component. These two approaches permit one, also, to learn the associated mathematics as it naturally unfolds in real world settings, in a relatively "painless" and jargon-free manner. The reader learns the mathematics required to consider the broad problem at hand, rather than learning mathematics according to the determination of a (perhaps) artificial curriculum. The underlying philosophy is different, as is the book format. The different format is required given the interactive and animated features that enliven the book and motivate the reader to explore diverse realms in the worlds of geography and mathematics and, especially, as they combine and interact.

1.2 Theory: Earth coordinate systems

Consider the Earth modeled as a sphere. The Earth is not a perfect sphere, but a sphere is a good approximation to its shape and the sphere is easy to work with using the classical mathematics of Euclid and others.

- Consider a sphere and a plane. There are only a few logical possibilities about the relationship between the plane and the sphere.
  - The sphere and the plane do not intersect.
  - The plane touches the sphere at exactly one point: The plane is tangent to the sphere.
  - The plane intersects the sphere (**Figure 1.1**).
    - And it does not pass through the center of the sphere: In that case, the circle of intersection is called a small circle (Plane *A*).
    - And it does pass through the center of the sphere: In that case, the circle of intersection is as large as possible and is called a great circle (Plane *B*).
- The planes that contain great circles are called diametral planes, because they contain a diameter of the sphere representing the Earth.
- Great circles are the lines along which distance is measured on a sphere: They are the geodesics (shortest-distance path between two points) on the sphere. **Figure 1.2** shows a route along a small circle

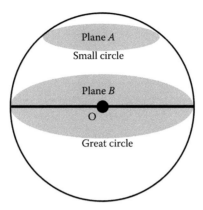

Figure 1.1 *A great circle on a sphere is formed by a plane (such as Plane B) cutting the sphere through the center, O, of the sphere. A small circle on a sphere is formed by a plane (such as Plane A) cutting into the sphere at a different location from the center, O, of the sphere.*

between two locations (40N, 90W) and (40N, 0) and also one along a great circle; the great circle route is the shorter route. What is a geodesic in the plane? The software used in the top image measured the great circle route (top line) as 4562 miles and it measured the small circle route (along the 40th parallel, bottom line) as 4786 miles.

- In the plane, the shortest distance between two points is measured along a line segment and is unique.
- On the sphere, the shortest distance between two points is measured along an arc of a great circle.
  - If the two points are not at opposite ends of a diameter of the sphere, then the shortest distance is unique.
  - If the two points are at opposite ends of a diameter of the sphere, then the shortest distance is not unique: One may traverse either half of a great circle. Diametrically opposed points are called antipodal points: Anti + pedes, opposite + feet, as in drilling through the center of the Earth to come out on the other side. The Earth's diameter separates pairs of antipodal points.
- To reference measurement on the Earth-sphere in a systematic manner, we introduce a coordinate system (**Figure 1.3**). A coordinate system requires a point of origin, a set of axis or reference lines, and a system of addressing any point within the system.
  - One set of reference lines is produced using a great circle in a unique position (bisecting the distance between the poles): The Equator. A set of evenly spaced planes, parallel to the equatorial plane, produces a set of evenly spaced small circles on the Earth's surface, commonly called parallels. They are called parallels because it is the planes that are parallel to each other (e.g., Planes *A* and *B* in **Figure 1.1**).

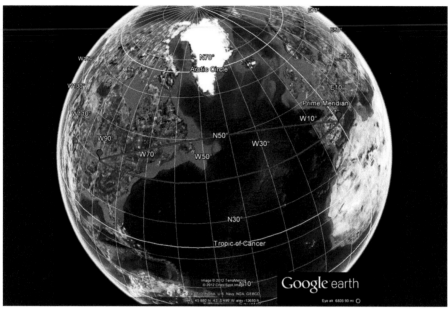

Figure 1.2 *The shortest distance between two points on a sphere is measured along an arc of a great circle. Two different views of the same paths, on the sphere and in the plane. Source of base maps: Based on Google Earth™ mapping service (top) © 2012 Google and Image © 2012 TerraMetrics, Data SIO NOAA US Navy NGA GEBCO, ©2012 Cnes/Spot Image; and, Great Circle Mapping Tool (bottom).*

- Another set of reference lines is produced using a half of a great circle, joining one pole to another, that has a unique position: The half of a great circle that passes through the Royal Observatory in Greenwich, England (three points determine a circle—both poles and Greenwich). The historical decision at the Meridian Conference in 1884 produced the uniqueness in selection. Choose a set of evenly spaced halves of great circles obtained by rotating

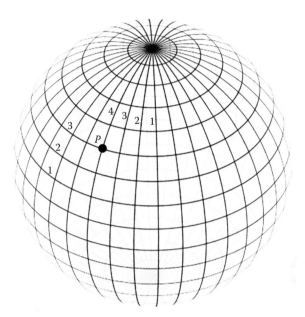

Figure 1.3 *Location of P is shown at three parallels north of the Equator and four meridians west of the Prime Meridian; it might equally be described, by doubling the number of parallels and meridians, as six parallels north and eight meridians west.*

the diametral equatorial plane about the polar axis of the Earth. These lines are called meridians: Meri + dies = half day, the situation of the Earth at the equinoxes. The unique line through Greenwich is called the Prime Meridian; other halves of great circles are simply meridians.

- This particular reference system for the Earth is not unique; an infinite number is possible. There is abstract similarity between this particular geometric arrangement of a coordinate system on a sphere and the geometric pattern of Cartesian coordinates in the plane.
  - To use this arrangement, one might describe the location of a point, *P*, on the Earth-sphere as being at the 3rd parallel north of the Equator and at the 4th meridian to the west of the Prime Meridian. This description locates *P* according to one reference system; however, someone else might employ a different reference system with a finer mesh (halving the distances between successive lines). For that person, a correct description of the location of *P* would be at the 6th parallel north of the Equator and at the 8th meridian to the west of the Prime Meridian. Indeed, an infinite number of locally correct designations might be given for the single point, *P*: An unsatisfactory situation in terms of being able to replicate results. The problem lies in the use of a relative, rather than an absolute, locational system (**Figure 1.3**).

- To convert this system to an absolute system, that is replicable, employ some commonly agreed-upon measurement strategy to standardize measurement. One such method is the assumption that there are 360 degrees of angular measure in a circle.
  - Thus, P might be described as lying 42 degrees north of the Equator, and 71 degrees west of the Prime Meridian. The degrees north are measured along a meridian; the degrees west are measured along the Equator or along a parallel (the one at 42 north is another natural choice). The north/south angular measure is called latitude; the east/west angular measure is called longitude.
  - The North and South poles are 90 degrees north latitude, and 90 degrees south latitude, respectively. The angle between the equatorial plane (Plane B in **Figure 1.1**) and the poles is 90 degrees. The polar axis is perpendicular to a diameter in the equatorial plane.
  - The use of standard circular measure creates a designation that is unique for P; at least unique to all whose mathematics rests on having 360 degrees in the circle.
  - Parts of degrees may be noted as minutes and seconds, or as decimal degrees. A degree (°) of latitude or longitude can be subdivided into 60 parts called minutes ('). Each minute can be further subdivided into 60 seconds ("). Thus, 42 degrees 30 minutes is the same as 42.50 degrees because 30/60 = 50/100. Current computerized mapping software often employs decimal degrees as a default; older printed maps may employ degrees, minutes, and seconds. Thus, the human mapper needs to take care to analyze the situation and make appropriate conversions prior to making measurements of position. Such conversion is simple to execute using a calculator. For example (see **Figure 1.4**), 42 degrees 21 minutes 30 seconds converts to $42 + 21/60 + 30/3600$ degrees = 42.358333 degrees; powers of ten replace powers of 60.
- **Figure 1.5** shows the described reference system placed on a sphere. What might be called a Cartesian grid in the plane is called a graticule on the sphere.
  - Points along one parallel all have the same latitude; they are the same distance above or below, north of or south of, the Equator.
  - Points along one meridian all have the same longitude; they are the same distance east or west of the Prime Meridian.
  - Spacing between successive parallels or meridians might be at any level of detail; however, when circular measure describes the position of these lines, that description is unique, according to the standard of using 360 degrees in a circle.
  - One spacing for the set of meridians that is convenient on maps of the world is to choose spacing of 15 degrees between successive

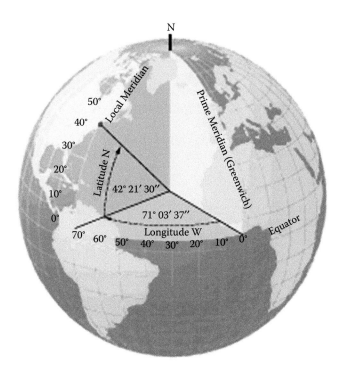

Figure 1.4 *Latitude and longitude as angular measures in relation to the center of the Earth-sphere. Source: Based on GeoSystems Global Corporation (later MapQuest) image, 1998.*

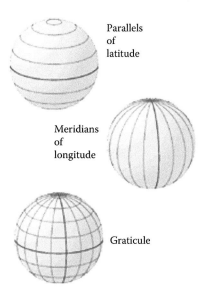

Figure 1.5 *Earth-sphere graticule composed of parallels and meridians. Source: Based on GeoSystems Global Corporation (later MapQuest) image, 1998.*

meridians. The reason for this is that since the meridians converge at the ends of the polar axis, it follows then that each meridian represents the passage of one hour of time—given the agreement to partition a day into 24 hours—24 times 15 is 360, so that meridians also mark time, at least in a general way and independent of political and social convention regarding awkward partitioning of urban areas, and so forth.

- Bounds of measurement (see **Figure 1.6**).
    - Latitude runs from 0° at the Equator to 90°N or 90°S at the poles.

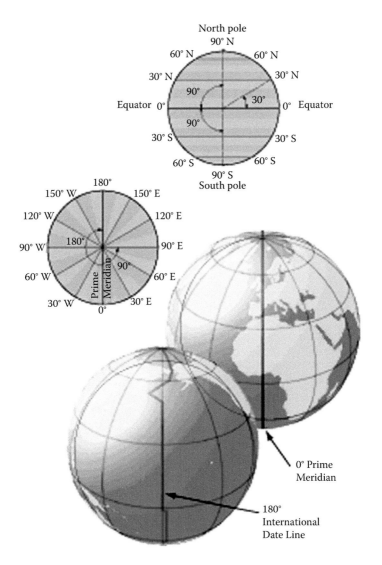

Figure 1.6 *Bounds of measurement for latitude and longitude. Source: Based on Geo-Systems Global Corporation (later MapQuest) image, 1998.*

- Longitude runs from 0° at the Prime Meridian to 180° east or west, halfway around the globe. The International Date Line follows the 180° meridian, making a few jogs to avoid cutting through land areas or island groups and thus separating adjacent people into two different days.

Latitude and longitude can also be understood by their connection to the Cartesian coordinate system. Picture a standard Cartesian coordinate system with an X axis and a Y axis (**Figure 1.7**). This system, when overlaid on the Earth, gives a framework for visualizing latitude and longitude. On the Earth, the X axis becomes the Equator, with a value of $Y = 0$, or zero degrees north or south latitude. The Y axis becomes the Prime Meridian, with a value of $X = 0$, or zero degrees east or west longitude.

Longitude values are the X values as one moves left (west) or right (east) along the Equator or along any other line of latitude north or south of the Equator. Latitude values are the Y values as one moves up (north) or down (south) along the Prime Meridian or along any other meridian. Confusion may arise because in most everyday speech when referring to "latitude, longitude", latitude is typically mentioned first, and it is tempting to think of these as being equivalent to x, y, with latitude being "x" and longitude being "y." However, it is important to remember that latitude is "y" and longitude is "x." And in location-enabled devices and tools, such as Global Positioning Systems (GPS) and Geographic Information Systems (GIS), latitude and longitude are entered as y and x, respectively.

The Cartesian coordinate system also helps us understand why the sign (positive or negative) of latitude and longitude is important. The Equator divides the area above the X axis, the northern hemisphere, from the area below the

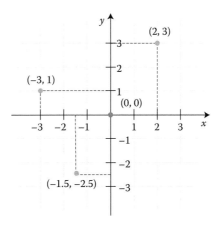

Figure 1.7 *Cartesian coordinate system. Source: Based on material from K. Bolino, 2008, released into the public domain, Wikipedia, http://en.wikipedia.org/wiki/ File:Cartesian-coordinate-system.svg*

X axis, the southern hemisphere. The Prime Meridian divides the area to the right of the *Y* axis, the eastern hemisphere, from the area to the left of the *Y* axis, the western hemisphere. Any *x* value to the right, or east, of the *Y* axis is positive. Therefore, any longitude in the eastern hemisphere is positive. Any *x* value to the left, or west, of the *Y* axis is negative. Therefore, any longitude in the western hemisphere is negative. Similarly, any *Y* value above, or north, of the *X* axis is positive. Therefore, any latitude in the northern hemisphere is positive. Any *Y* value below, or south, of the *X* axis is negative. Therefore, any latitude in the southern hemisphere is negative.

Thus, given the following coordinate pairs, one can determine their correct hemisphere:

X, Y	Eastern and northern hemisphere
–X, Y	Western and northern hemisphere
X, –Y	Eastern and southern hemisphere
–X, –Y	Western and southern hemisphere

Therefore, all of the United States is in the northern hemisphere and is composed of positive *Y* values in terms of latitude. Most of the United States is in the western hemisphere, and is described with negative *X* values. The only exception is the far western part of the Aleutian Islands, which extends into the eastern hemisphere.

1.3 Theory: Earth's seasons—A visual display

The primary cause of Earth's seasons is the tilt of the Earth's polar axis in relation to the plane of the ecliptic (Sun's equatorial plane). The tilt is 23.5 degrees and this tilt has far-reaching implications. **Figure 1.8** illustrates the annual revolution of the Earth around the Sun on the plane of the ecliptic. Note the constant tilt of the Earth's polar axis. The circle of illumination, or terminator, marks the separation of day and night. At the equinoxes (about March 21 and September 21), the circle of illumination passes through both poles (the only time of the year at which it does so). Thus, at those times, the circle of illumination bisects all parallels of latitude; half of the day is on the dark side, half of the day is on the light side for all latitudes. Hence, "equinox = equal night" occurs all over the world. On or about June 21, the northern half of the Earth receives the most extra sunlight that it will all year–due to the axis tilt. Thus, it is the beginning of summer (relatively warm season) in the northern hemisphere. It follows, therefore, that the southern half of the Earth on or about June 21 receives less sun than on the equinox. Thus, it is winter in the southern hemisphere beginning on or about June 21. Similar arguments apply on or about December 21.

As shown in **Figure 1.8**, the North Pole is in constant darkness on December 21. Hence, it is the beginning of winter in the north and consequently summer in the south, on that date. The summer and winter positions are called

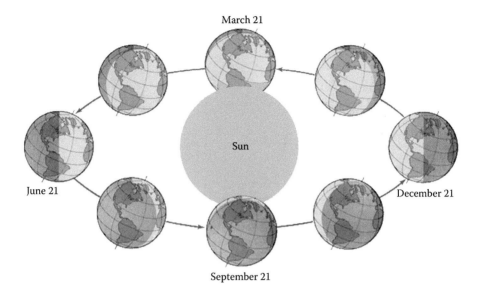

Figure 1.8 *Annual revolution of the Earth around the Sun. Source: Based on an image in Kolars, John F. and Nystuen, John D., Human Geography: Spatial Design in World Society, Englewood Cliffs: McGraw-Hill, 1974. Original drawings by Derwin Bell.*

solstices or "sun stops" because for a few days, depending on the hemisphere, and whether it is the summer or winter solstice, the sun stops getting higher in the sky with each successive solar noon or stops getting lower in the sky (depending on hemisphere).

Figure 1.9 shows a closer view of the detail of the tilt of the Earth's polar axis in relation to the plane of the ecliptic. Latitude and longitude measure

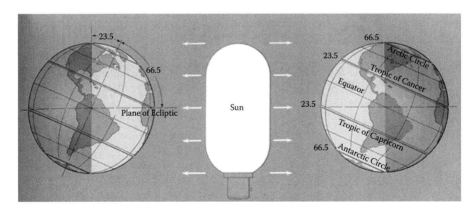

Figure 1.9 *Angular measure associated with tilt of the Earth's polar axis. Source: Based on an image in Kolars, John F. and Nystuen, John D., Human Geography: Spatial Design in World Society, Englewood Cliffs: McGraw-Hill, 1974. Original drawings by Derwin Bell.*

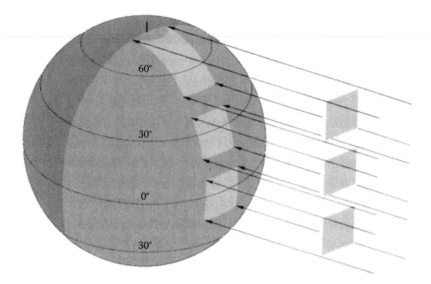

Figure 1.10 *Sunlight intensity is diminished as one travels toward the Earth's poles.*
Source: Based on an image in Kolars, John F. and Nystuen, John D., Human
Geography: Spatial Design in World Society, Englewood Cliffs: McGraw-Hill, 1974.
Original drawings by Derwin Bell.

the position of the Sun's rays. The ray of the Sun is directly overhead at 23.5 degrees north on June 21 (as far north as it ever will be); at 23.5 degrees south on December 21 (as far south as it ever will be); and at zero degrees on the equinoxes.

In addition to this basic geometry, travelers often note, when journeying closer to the Earth's Equator than they are accustomed to doing, that the Sun seems "hotter" than they were expecting. That observation is not surprising and it is based upon simple geometry. Sunlight intensity at the surface of the Earth is a function of latitude. **Figure 1.10** shows why: At higher latitudes, sunlight is spread out over a wider area than it is closer to the Equator and is thus less intense. The Sun is most intense when the direct ray is overhead or orthogonal to a tangent plane to the Earth. The even poleward spreading is a function of the relatively constant curvature of the Earth's surface.

1.4 Theory: Precision of latitude and longitude values

As we have seen, latitude and longitude values can be subdivided into units smaller than a degree. This subdivision is useful to denote locations on the Earth. As the number of significant digits to the right of the decimal point increases, one can pinpoint locations with increasing precision. One way to conceptualize this idea is that at the Equator, the length of one degree of longitude is 111.2 kilometers (km), or 69 miles (mi). In **Chapter 2**, we will explain how this figure is calculated, but for now, the length of a degree can help us understand the value of precision in measuring on the Earth. Given that one degree is 111.2 km, then the length of one minute is 111.2/60, or 1.853 km, and the length of one second is 1.853/60 = 0.03089 km, or 30.89 meters (m). If any given set of coordinates is off by one second, then the distance off on the Earth's surface is just over 30 meters. And remember, this is at the Equator: An error of one second elsewhere on the Earth's surface is a smaller distance but still merits attention.

Let us explore these same measurements using decimal degree format instead of degrees, minutes, seconds. If one degree is 111.2 km at the Equator, then 0.1 degree is 11.2 km, 0.01 degree is 1.2 km, 0.001 degree is 0.12 km, 0.0001 degree is 0.012 km, or 12 meters, and 0.00001 degree is 0.0012 km, or 1.2 meters. Therefore, four significant digits to the right of the decimal will pinpoint a location to within 12 meters, and five significant digits to the right of the decimal will pinpoint a location to within 1.2 meters on the Earth's surface. Therefore, keeping all of the significant digits in any measurement is important! Earth measurement is definitely not a case when numbers should be truncated or rounded.

It is important to note, however, that simply because Global Positioning Systems (GPS) receivers and GPS apps on smartphones can provide very precise locations, coupled with Geographic Information Systems (GIS)-based digital maps that allow for zooming into very detailed scales, one must use caution in the confidence placed on the resulting obtained locations. A large number of significant digits to the right of a decimal for any latitude and longitude coordinate pair does not necessarily mean that the position was gathered from equipment or a sensor with a high degree of accuracy. Precision is not the same as accuracy. Having the significant digits may just be a capability of your mapping software or GPS rather than an indication of the accuracy of your measurement.

Some readers might consider the words "accuracy" and "precision" to be synonyms. It is important to clarify the distinction between these words as we will refer to it throughout this book. "Accuracy" refers to whether measurements taken within a system are close to an accepted value (Math Is Fun, 2012). The closer the measurements are to an accepted value, the more accurate that set of measurements is said to be. "Precision," on the other hand, refers to whether measurements taken within a system are close to each other. One way to model the differences is by using a target, where arrows shot at a target represent measurements within a system. **Figure 1.11** shows images representing high accuracy and high precision (**Figure 1.11a**), high accuracy but low precision (**Figure 1.11b**), high precision but low accuracy (**Figure 1.11c**), and low accuracy and low precision (**Figure 1.11d**). In the best of all possible worlds, of course, one would wish for "high" in both categories. Mathematics may offer high accuracy; science demands replication of results and therefore high precision. This distinction is important both in terms of knowing your hardware, such as GPS units, as well as in terms of knowing your data.

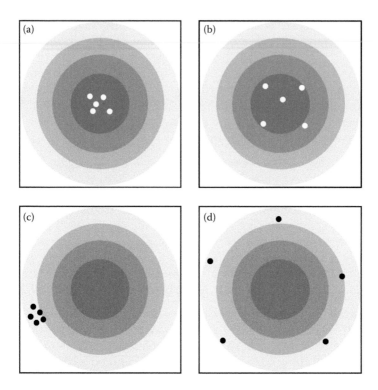

Figure 1.11 *All possible permutations of high and low, accuracy and precision. (a) High accuracy and high precision. (b) High accuracy but low precision. (c) High precision but low accuracy. (d) Low precision and low accuracy.*

As an example of knowing your data's accuracy and precision, consider a set of coordinates provided by the United States Geological Survey (USGS) for earthquake locations for the last 30 days for the globe. Earthquakes do not typically occur conveniently directly under seismic stations. In addition, a seismic station has not been constructed over every square meter of the Earth's surface. Therefore, earthquake epicenters are determined based on triangulation of seismic waves received from a network of seismic stations spaced around the world (**Figure 1.12**).

Two types of seismic waves are received at different times by different seismic stations, and from those times, a distance is determined from each station. This results in an estimate of the epicenter's location, reflected by the radii of circles centered on *A*, *B*, and *C* representing seismic stations in **Figure 1.12**. The word "estimate" is important. In fact, these epicenters are called "preliminary determinations of epicenters" to indicate the fact that they are mathematically calculated, as is *E* the intersection point in **Figure 1.12**, and not surveyed with instruments. An error in the measurement of the time the seismic waves were received by any one station would result in the size of the corresponding circle to be altered, and therefore, a compromising of

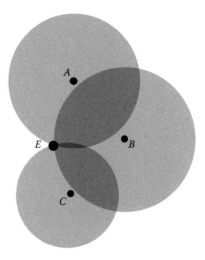

Figure 1.12 *A, B, and C represent seismic stations. The point E, calculated from circle intersections, represents the estimated epicenter location.*

the accuracy of the position of the epicenter. The coordinates may be quite precise, such as to six significant digits to the right of the decimal point, but the accuracy may not reflect this precision. Furthermore, the error produced by the misreading of the timing of the seismic waves may be random, resulting in a lower precision in terms of the location of any aftershocks recorded. Repeated measurements may not show the same results. The data are still valuable, but like any data, they must be as fully understood as possible so that their limitations can be recognized enabling the data user to make well-informed choices.

A location generated using a Global Positioning System (GPS) is derived in a similar manner. It offers another example of limitations that must be considered when using data. Each GPS satellite generates a signal that is received by the receiver held by the operator on the surface of the Earth. The receiver uses the messages it receives to determine the transit time of each message and computes the distances to each satellite based on the formula: Distance = rate × time. Because the signals travel at the speed of light, and because the GPS satellites travel in near-circular orbits with radii at 26,000 kilometers, the time it takes for the signals to travel from satellite to receiver is only a tiny fraction of a second. Therefore, timing is everything. Determining the precise time allows for precise positions to be calculated. The major difference between the determination of seismic epicenters and the determination of a ground position via GPS is that the former works in two dimensions while the latter works in three dimensions. Therefore, in the GPS world, instead of circles, we are working with spheres (**Figure 1.13**).

If only one satellite's signal is sensed by the GPS receiver, all the GPS receiver knows is that it is somewhere on a sphere at a certain distance from that

Figure 1.13 A general satellite configuration of six orbits, four satellite slots in each orbit. Source of image: NOAA, http://www.gps.gov/systems/gps/space/. Link, using the QR code, to an animation to see how satellite configurations change. Source of link embedded in QR code: Public domain, released by author El pak, 2007. http://en.wikipedia.org/wiki/File:ConstellationGPS.gif

satellite. If two satellites are sensed, then the GPS receiver knows it is at one of two points where the two spheres intersect. If three satellites are sensed, then the GPS receiver can determine the position on the ground typically within 10 meters of horizontal accuracy. If four or more satellites are sensed, a recreational GPS receiver can determine a position within three or four meters, and with mathematical differential correction, a high end professional receiver can pinpoint the user's location to within centimeters. Vertical accuracy is usually an order of magnitude coarser than the horizontal accuracy, so if the horizontal accuracy is within three meters, the vertical accuracy in terms of the elevation of the user may only be within 30 meters of the true elevation.

Three satellites might seem enough to solve for position since space has three dimensions and a position near the Earth's surface can be assumed. However, even a very small clock error multiplied by the very large speed of light results in a large positional error. Furthermore, GPS signals do not travel through most solid objects, and therefore, accuracy will suffer if the user is standing in heavy tree cover or in or near tall buildings. In smartphones, most of which have GPS capability, the position from GPS is enhanced by triangulation off of nearby cell phone towers. Of course, if no cell phone towers are nearby, the positional accuracy on a smartphone will be compromised.

1.5 Other Earth models

The discussion in this chapter has been focused on a spherical Earth. However, due to the centripetal force of the Earth's rotation, the Earth is not a perfect sphere. It is an oblate spheroid, with an equatorial diameter that is 42.72 kilometers (26.54 miles) longer than its diameter measured through the poles. Because we do not have a tape measure that we can wrap around the Earth to determine its exact shape and measurements, these all have to be mathematically calculated. Geodesy is the science that studies the size and shape of the Earth and its gravitational field. Further complicating the situation is that the shape, rotation, and revolution of the Earth changes over time.

When modeling the Earth in geography and in mathematics, three concepts must be simultaneously addressed: The physical Earth, the geoid, and the reference ellipsoid. The physical Earth is the actual shape and size of the planet; it is what we try to measure and model. Gravity is not even across the Earth's surface, due to the Earth's rotation, altitude, and differences in rock density. The geoid represents a surface of constant gravity. This is idealized by the notion of mean sea level, which is a theoretical level of the average height of the ocean's surface. It does not, however, correspond to the actual level of the sea at any given point at any instant in time, but is the halfway point between mean high and low tides. The geoidal surface is irregular, but smoother than the physical Earth's surface. The reference ellipsoid is a mathematically defined shape that is a "best-fit" to the geoid, most often for a continent, but potentially for the entire globe. One way to view goodness-of-fit is to consider the degree of separation between in the geoid and ellipsoid (the value N in **Figure 1.14**). The reference ellipsoid is smoother than the geoid, and it is the surface upon which coordinate systems are defined (**Figure 1.14**).

Another fundamental concept in spatial mathematics is the concept of a datum. We have already discussed the importance of a zero point when calculating horizontal position such as latitude and longitude. This concept is also critical with respect to other common mapping coordinate systems (such as Universal Transverse Mercator (UTM) coordinates, or state plane coordinates). However, where is the "zero point" in terms of elevation, or the "z" coordinate, given the complications in the shape of the Earth as described above? Datums help solve this problem. Datums are reference surfaces; that is, mathematically derived, against which position measurements are made. They define the location of zero on the measurement scale. Datums form the basis of coordinate systems. Hundreds of locally developed reference datums exist around the world usually referenced to convenient local reference points. Modern datums are based on increasingly accurate measurements of the shape of the Earth, however, and tend to cover larger

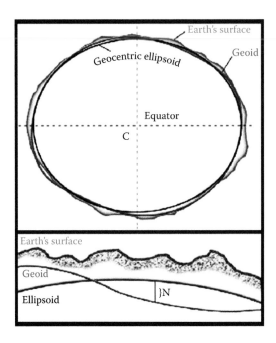

Figure 1.14 *Relationship of the Earth's surface, the geoid, and a reference geocentric ellipsoid. The height difference (N) between the geoid and the reference geocentric ellipsoid is the geoid separation. Source: Images of Kevin McMaster, URS Corporation; published here with permission from McMaster.*

areas. Horizontal datums are used to determine locations in latitude and longitude, while vertical datums are used to determine elevations above and depths below a reference surface.

We have discussed how the Equator and the Prime Meridian are used for the zero coordinates for latitude and longitude. However, for vertical measurements, three fundamentally different approaches exist. The choice of an appropriate datum depends on such factors as whether one is interested primarily in the height of land (topography) or the depth of water (bathymetry). Datums are updated as the surface of the Earth changes. They are also updated as measurement becomes increasingly accurate. For example, the North American Datum of 1927 was used for decades, such as on USGS topographic maps, and the digital maps that were derived from those paper maps, until it was revised in 1983. Similarly, the North America Vertical Datum of 1929 was in wide use until it too was revised in 1988. Most GPS receivers use, as their default, the World Geodetic System of 1984 as their datum. It is important to know what datum the map you are using was derived from, because every single x, y, and z coordinate on those maps is only valid for its specific datum. Converting from one datum to another across map edges or across time is necessary when working with maps at multiple scales and across large regions. We will return to the topic of datums when we address elevation data in a later chapter.

In the following section, you will have the opportunity to put the concepts of absolute location, latitude and longitude, precision, accuracy, and GPS into practice. In the first activity, you will use an online mapping service to determine the positions on the opposite side of the Earth, the antipodal points. You will also use a latitude-longitude finder to determine the position opposite from your current location, and you will examine land areas on the opposite side of the Earth using two antipodal maps.

1.6 Practice using selected concepts from this chapter

In this section, you will have the opportunity to practice with routes, coordinates, and precision and accuracy to enhance your understanding of the concepts presented in this chapter.

1.6.1 Antipodal points

Start a web browser and access Google Maps (http://maps.google.com). Access the "Earth" view to start Google Earth with its 3-D capabilities in your browser. If you do not see the "Earth" choice, install the Google Earth plug in from www. google.com. In the search box, enter the latitude and longitude of London as follows: 51.5, 0. Antipodal points are a pair of diametrically opposed points on a great circle; they are 180 degrees apart. The antipodal component to latitude is the negative of that latitude. The antipodal component to a western hemisphere longitude is that longitude plus 180. The antipodal value to an eastern hemisphere longitude is that longitude minus 180. Thus, the antipodal component to 51.5 degrees (north) latitude would be −51.5 degrees (south). The antipodal component to zero degrees longitude is 0 + 180, or 180 degrees longitude. The antipodal component to 40 degrees north latitude and 105 degrees west longitude is 40 degrees south latitude and −105 + 180 = 75 degrees east longitude. Enter the latitude, longitude value of 45, −90 into the search box in Google maps. In what region of the world are you located? You should be in Wisconsin in the United States. Next, enter −45, −72. You should be in the country of Chile. Calculate the antipodal point. Enter the coordinates of the antipodal point. You should be in Mongolia. Find the latitude and longitude of your own community.

One tool that enables you to do this task is the Esri EdCommunity latitude-longitude finder. After you calculate it, enter that latitude and longitude coordinate pair into Google maps. Then, calculate the antipodal point, and enter those coordinates. What location is directly opposite your community on the globe? http://edcommunity.esri.com/maps/geocoder/AGS_EdComm_Geocoder.html. Answers will vary; for example, the antipodal point to New Orleans at 30 North 90 West is 30 South 90 East, which is west-northwest of Perth, Australia, in the Indian Ocean.

 To check your answers and visualize locations on the opposite side of the globe, access the Antipode Map, on http://www.antipodemap.com.

 The phrase "dig to China" among Americans reflects the erroneous belief by some that if you could dig a hole from your current position through the center of the Earth, eventually you would emerge in China. Is this true? To visualize land masses on the opposite side of the globe, visit "other side of the world" map on: http://www.peakbagger.com/pbgeog/worldrev.aspx.

Why is it impossible for China to be on the opposite side of the world from the USA? What is on the opposite side of the world from any point in the 48 contiguous United States? Is it a land mass or an ocean? Identify the land mass or ocean that is at this location. You will find that on the opposite side of the world from most of the continental United States is not China, but the Indian Ocean. In which countries would a hole in a straight line emerge in China? Most of Argentina and Chile would meet this criterion. What island country is directly opposite to sections of Spain and Portugal? New Zealand meets this criterion. Check early references on problems associated with the simultaneous global mapping of an infinite number of pairs of antipodal points (Tobler, 1961a; Arlinghaus and Nystuen, 1985).

In the next activity, you will test the concepts of accuracy and precision using the GPS function in an ordinary smartphone. You will also have the opportunity to save your recorded path on your smartphone and map and analyze your path in a web-based GIS.

1.6.2 Capturing points with a smartphone

On an iPhone or Android smartphone, enable the GPS or "location" function, and then access whatever default mapping tool you may have on the smartphone. Go outside and plot your current position on the map with your phone's default mapping tools. If possible, switch the base map to a satellite image and zoom in as much as you can. Test your positional accuracy in an open area versus in a wooded area or near a building. Can you determine any differences in the horizontal accuracy? On most smartphones, the circle indicating your accuracy should become larger in areas of heavy timber or buildings, indicating that the accuracy is degraded in such situations. The latitude and longitude readings will also fluctuate more rapidly in such places.

Place your hand across the top of the phone. Does your accuracy degrade, and by how much? Depending on your local conditions, your signal should degrade from a few meters to tens or even hundreds of meters or stop working entirely. Start walking. How accurate is your position relative to the base map? Your recorded position on the base map should fall onto the opposite side of a nearby building, or in the next block, or on the other side of town. Go inside. Can your phone still determine its position? If so, does its accuracy degrade in any way, and by how much? When you first enter a building, your position should be captured for a while. If you have many nearby windows, or if you are under a certain kind of roof, your smartphone may be able to detect your position for a while, or even indefinitely, though with a reduced accuracy and precision as when you were outside.

To analyze the positions that you collect on your smartphone in a more rigorous way, collect a track on your phone that you can map later using an online GIS. On an iPhone, use Motion X GPS or an equivalent tool to capture coordinates as a track. On an Android, use MyTracks as an equivalent app to accomplish the same thing. Start your collection while inside a building, and walk outside while continuing to collect tracks. As you did in the previous activity, cover part of the phone with your hand, and access the Motion X GPS or MyTracks screen while doing so. Can you determine how many GPS satellites you are blocking? Depending on your local environment, and the version of the smartphone app that you are using, you should be able to see that you are blocking at least one and possibly all of the GPS satellites.

Save your track as a GPX file. A GPX file is a "GPS Exchange Format" file that contains coordinates that you collect in the field. Email the GPX track to yourself using the function in Motion X GPS or MyTracks. Retrieve your email and save your GPX file to your local computer. Or, alternatively, save your track as a KML file and email it to yourself. A KML file is a Keyhole Markup Language file in wide use for Earth browsers such as Google Earth and other tools.

Now that you have collected your track, you are ready to map it. To do so, access ArcGIS Online (www.arcgis.com/home) and make a new map. Use the Add button to find your GPX or KML file stored on your local computer or online. Once your track is mapped, change the symbols of your track points if desired, to map your tracks based on elevation or time. Change your base map to a satellite image so that you can determine the accuracy of your track based on the image beneath your points. Remember that the image base map is not perfect either—it was created using a series of mathematical algorithms and contains distortion and error. Use the measure tool to determine the distance between where you know you walked on the ground, using the image to verify, versus where the track points are located.

How different are your track points from where you believe you walked based on the satellite image, on average? You should be able to determine that your points, with a strong GPS signal, are within a few meters of where you actually walked. What is the farthest difference between your track points and your satellite image? Depending on your signal strength, it may be several to even tens of meters of difference. What distance represents the closest match between where you believe you walked versus the location of the track points? If your signal was strong, you might be able to find a point that is only one meter off from the true point where you stood on the Earth. How did the accuracy of the points gathered when you were inside match your true location? How did the points gathered when you first stepped outside match your true location versus as you were walking later on? The accuracy gradually increases the longer you are outside, and the farther you are from buildings. Why does the accuracy of your points generally improve as you remain outside? The accuracy of your points improves because the GPS signals are in "clear view" of the smartphone. In the absence of GPS signal, most smartphones attempt to triangulate and determine your position based on nearby cellphone towers. This triangulation, too, is heavily dependent upon mathematics.

In the next section, you will have the opportunity to map a track collected with a smartphone that one of the authors collected. This track has been symbolized based on the time that the data were collected. The track contains teachable moments involving GPS, position, precision, and accuracy.

Analyze the following map that was created by one of the authors of this book (Kerski) using Motion X GPS on an iPhone and mapped using ArcGIS Online. Access www.arcgis.com/home (ArcGIS Online) and search for the map entitled "Kendrick Reservoir Motion X GPS Track" or directly to http://bit.ly/Rx2qVp (**Figure 1.15**). Open the map. This map shows a track that the author took around Kendrick Reservoir in Colorado. This map was symbolized on the time of GPS collection, from yellow to gradually darker blue dots. Note the components of the track to the northwest of the reservoir. These pieces were generated when the smartphone was just turned on and the track first began, indicated by their yellow color. They are erroneous segments and track points. How do you know this? You know the author walked around Kendrick Reservoir, but there is another way to determine that this is an erroneous track. Notice how the track cuts across the terrain and does not follow city streets or sidewalks. Change the base map to a satellite image base. Cutting across lots would not have been possible on foot given the fences and houses obstructing the path. When the author first turned on the iPhone, not many GPS satellites were in view of the phone. As the author kept walking and remained outside, gradually, the phone could sense a greater number of GPS satellites, and as the number of satellites increased, the triangulation used more spheres, and the positional accuracy improved until the track points mapped closely represented the author's true position on the Earth's surface.

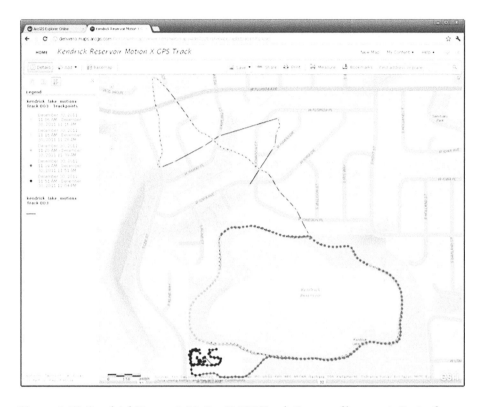

Figure 1.15 *Kendrick Reservoir Motion X GPS Track. Source of base map: Esri software.*

Use the distance tool to answer the following question: How far were the farthest erroneous pieces from the lake? Although it depends on where you measure from, some of the farthest erroneous pieces were 600 meters from the lake. Click on each dot to access the date and time each track point was collected. How long did the erroneous components last? Again, it depends on which points you access, but the erroneous components lasted about 10 minutes. At what time did the erroneous track begin correctly following Kerski's walk around the lake? This occurred at 11:12 a.m. on the day of the walk.

How many laps around the lake did the author walk? The author walked three laps. How far apart are the track points, on average? The track points are about 10 meters apart. How much time elapses, on average, between each point? About 10 points were collected each minute, so the time between each point is about six seconds. As is visible on the map, as time went on, the smartphone did a better job of locking onto the available GPS satellites, to the point where the author was able to draw some letters to the southwest of the reservoir. Zoom to these letters. What letters was the author trying to draw? How high is each letter? The author was trying to draw the word "GIS," and each letter is about 42 meters "high." How does the spatial accuracy of the smartphone and the GPS signals it uses affect how large the letters have to be in order to draw? The letters have to be at least this high for the shapes to be discernible. If they are anything smaller, the GPS would not be able to distinguish the angles of the letters necessary to read them.

1.6.3 Great circle routes

In this section, you will have the opportunity to practice with great circle routes using a web-based GIS and with a GPS receiver or smartphone. You will start by measuring distances and thinking about great circle routes as you do so.

First, use a web-based GIS, ArcGIS Online, (www.arcgis.com/home) that you used in the above activity. Start a new map by selecting "map." Zoom out until you can see London, England, and Los Angeles, California. Access the measure tool and click on the measure line symbol. Set the distance units to miles or kilometers—whichever you prefer. Click on Los Angeles and then click on London, as shown below (**Figure 1.16**).

Why is the shortest line distance from Los Angeles to London shown as an arc? What is another name for this arc? The shortest distance is shown as an arc because the Earth is a sphere, and the shortest distance between two points is an arc on that sphere, known as a great circle. What countries and land masses does your arc touch, and therefore, which countries would you see if you were on an airline flight between the two cities? You pass over Canada,

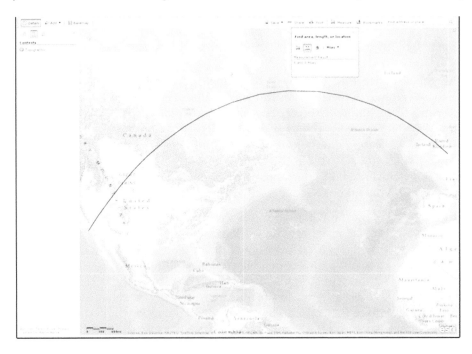

Figure 1.16 *The shortest distance from Los Angeles to London. Source of base map: Esri software.*

including northern Quebec, Greenland, Northern Ireland, Wales, and England. What did you determine the distance to be between the two cities? Repeat the process twice more and average your three measurements. Why do you think the distance is slightly different each time you measure it? The distance is approximately 5500 miles, but is different between measurements because of the challenge of placing the mouse exactly on the same two points each time.

Zoom in to a larger scale map. How long does any distance in this application have to be before it begins to look like an arc? Your answer may vary, but lines begin to look like an arc when they become longer than 1000 miles.

Measure the distance along the Equator, from northern Ecuador to western Kenya. What shape is your drawn line? Why does it have this shape? A line along the Equator in this mapping application will look straight because the Equator is the great circle route and a straight line in the Web Auxiliary Mercator projection that this map is in. Zoom to Tromvik, Norway, and then to Barrow, Alaska, and measure the distance between the two towns. Why does the line have this shape? The line has a shape like the letter "U" because of the map projection. Which hemispheres would you likely travel through in order to fly the shortest straight line distance between the two towns? You will travel through the northern, eastern and western hemispheres on your journey. To better visualize your route, you will need a more suitable map projection, which we will discuss in a later chapter.

Another way to visualize the great circle routes on the Earth-sphere is with a GPS receiver or smartphone. Most standard GPS receivers have a "Go To" function where the user can input a set of coordinates and ask the GPS to route the user to those coordinates. Because this route is the shortest distance between the current location and the desired location, it will be the great circle route. On any GPS receiver, or on a smartphone with a GPS app, enter the following coordinates for London, England: 51.5171 degrees north latitude, 0.1062 degrees west longitude. Next, access the "Go To" function to determine the distance and direction from your current location to the location that you have just entered. How far is London from your current location? What direction is it from where you are located? This depends on your location, but London could be at least 4000 miles away. From the United States, does the route take you significantly north of due east? Why? Does the route correspond to the great circle route that you mapped above with ArcGIS Online? The route does take you significantly north of due east, again, because of the spherical Earth and the great circle route. Start walking toward London, following the direction indicated on the GPS receiver. What is your current speed? Calculate how long it would take you to reach London if you could walk all the way there at your current speed. Calculate the day it would be in London when you arrived. If you walked four miles per hour and London was 4000 miles away, to reach London would require 1000 hours, or 41.6 days. However, you may find yourself in need of a boat!

In the next section, you will use a web GIS mapping service to examine specific locations around the planet.

1.6.4 Latitude and longitude, hemispheres, and precision

Start a web browser session and access ArcGIS Explorer Online (http://explorer. arcgis.com). Start a new map. Change the base map to National Geographic. In the search box, enter 39.58841, 105.64335 for the latitude and longitude, respectively. As indicated earlier, some GIS tools such as ArcGIS Online force coordinate entry as latitude, longitude (y, x) rather than longitude, latitude (x, y). Zoom out until you can recognize some landforms and countries. What country are you in? Recall the discussion in this chapter about latitude and longitude and the Cartesian coordinate system. You are in the northern and eastern hemispheres because both the latitude and longitude values are positive. As you zoom out to a smaller scale, you will see that you are in China. In the search box, now enter −39.58841, 105.64335, making sure this time that the latitude is negative. In what hemispheres is this location, and why? In what ocean is this location? What is the nearest continent? Use the measurement tools to measure the distance to it, indicating the units you are using and the cardinal direction in which it lies in relation to the point. You are now in the southern and eastern hemispheres, because your latitude is negative and your longitude is positive, in the Indian Ocean, about 630 miles southwest of Australia.

Next, in the search box, enter −39.58841, −105.64335, making sure this time that both the latitude and the longitude are negative. In what hemispheres is this location, and why? As you did with the previous location, indicate the nearest continent, the distance to it, and its cardinal direction. Change the base map to topographic. Now you are in the southern and western hemispheres, because your latitude and longitude are both negative, putting you in the lower left of the Cartesian coordinate system as cast on the world. Specifically, you are in the Pacific Ocean, about 1,700 miles due west of Chile.

Next, in the search box, enter 39.58841, −105.64335, making sure this time that the latitude is positive and the longitude is negative. This time, add your point to map notes (using the plus sign on the location pop-up to pull up a menu with "add point" as an option), and a push pin should remain in this location. What mountain is the map centered on, and what is its elevation? Zoom out until you see the state and country. What state and country is this mountain in? In what hemispheres is this location, and why? This is Mount Evans, Colorado, USA, and is located in the northern and western hemispheres because its latitude is positive and its longitude is negative.

Now let us visualize how precision manifests itself on maps and why it matters. Zoom back to 39.58841, −105.64335. Use the search box to search for 39.5884, −105.6433, thus removing one significant digit off of each of the latitude and longitude values. Add your point to the map notes. How far is this point from the summit of the mountain? In what direction is this point from the summit of the mountain, and why? Remove one more digit and search for 39.588, −105.643, again adding it to the map notes. How far is this point from the summit of the mountain? In what direction is this point from the summit of the mountain, and why? Remove one more digit and search for

39.58, −105.64, again adding it to the map notes. How far is this point from the summit of the mountain? In what direction is this point from the summit of the mountain, and why? Remove one more digit and search for 39.5, −105.6, again adding it to the map notes. How far is this point from the summit of the mountain? In what direction is this point from the summit of the mountain, and why? Finally, remove one more digit and search for 39, −105, again adding it to the map notes. How far is this point from the summit of the mountain? In what direction is this point from the summit of the mountain, and why? The points should be getting gradually farther from the mountain peak, because you are removing significant digits from the coordinate pair.

Examine all five of your points. In what "direction" did the points "move" as you cut significant digits off of your values? The points move to the south and to the east, toward the origin of the Cartesian coordinate System, which on the Earth is 0,0, or in the Atlantic Ocean off the coast of western Africa. In what direction will your points move if you round up to the next significant digit? Test your answers on your map. The points will move north and west if you round "up" to the next digit.

Why, then, is keeping significant digits to the right of the decimal point important when you are mapping your data? This is important because a loss of even one significant digit changes the location of any mapped point. **Figure 1.17** illustrates this situation in which the hundredths decimal place is dropped.

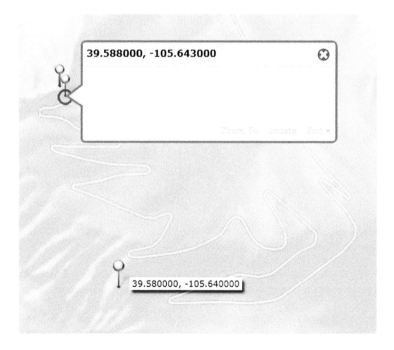

Figure 1.17 *Dropping the hundredths decimal place makes a noticeable difference in geographic location.*

Name three issues or problems that you might want to address where you would want to keep all of these significant digits. These could include locating natural gas pipelines for repair, the location of current political unrest in a city, or community water wells. Conversely, there may be issues or problems where you do not need to keep all significant digits. Name three issues or problems where, say, only two significant digits to the right of the decimal point would be sufficient. These could include earthquake epicenters, the location of the center of a high pressure system, or the ideal field area to study a particular invasive plant species. As we said in the theory section, rounding or truncation of decimal values can cause substantial problems in real-world settings!

1.6.5 Final considerations

How has your perspective on the ways that geography and mathematics intersect changed after reading this chapter? Do you think that latitude and longitude will be used by an increasing number of tools and applications in the future? How can you use the concepts and activities presented in this chapter in your own studies, teaching, and research?

1.7 Related theory and practice: Access through QR codes

Theory

Persistent archive:
University of Michigan Library Deep Blue: http://deepblue.lib.umich.edu/handle/2027.42/58219
From Institute of Mathematical Geography site: http://www.imagenet.org/

Arlinghaus, S. L. and J. D. Nystuen. 1986. *Mathematical Geography and Global Art: The Mathematics of David Barr's 'Four Corners Project.'* Institute of Mathematical Geography. Scroll down to find antipodal landmass map is near the end of the book. http://www-personal.umich.edu/~copyrght/image/monog01/fulltext.pdf
Arlinghaus, S. L. Parallels Between Parallels. 1990. *Solstice: An Electronic Journal of Geography and Mathematics.* Volume I, No. 2. Scroll down to find this article in eBook Monograph 13. http://www-personal.umich.edu/~copyrght/image/solstice/sols290.html
Arlinghaus, S. L. and W. C. Arlinghaus. 2005. *Spatial Synthesis: Centrality and Hierarchy.* Volume I, Book 1. Ann Arbor: Institute of Mathematical Geography. http://www.imagenet.org http://www-personal.umich.edu/~copyrght/image/books/Spatial%20Synthesis2/1ndex.htm

Practice

From Esri site: http://edcommunity.esri.com/arclessons/arclessons.cfm

Kerski, J. Analyzing Coordinates. Complexity level: 1, 2. http://edcommunity.esri.com/arclessons/lesson.cfm?id=488
Kerski, J. Analyzing Coordinates. Complexity level: 3a. http://edcommunity.esri.com/arclessons/lesson.cfm?id=494
Kerski, J. Analyzing Coordinates. Complexity level: 3b. http://edcommunity.esri.com/arclessons/lesson.cfm?id=520
Kerski, J. Analyzing Coordinates. Complexity level: 3c. http://edcommunity.esri.com/arclessons/lesson.cfm?id=500
Kerski, J. The Land of Cartesia: Coordinates. Complexity level: 3. http://edcommunity.esri.com/arclessons/lesson.
  cfm?id=513
Kerski, J. The Land of Cartesia ArcGIS version. Complexity level: 3. http://edcommunity.esri.com/arclessons/lesson.
  cfm?id=525
Kerski, J. Mapping and Analyzing Coordinates. Complexity level: 3, 4, 5. http://edcommunity.esri.com/arclessons/lesson.
  cfm?id=557
Kerski, J. Video Drawing With GPS, Mapping with GIS. http://youtu.be/cfmQe5OLu0A

2

Location, Trigonometry, and Measurement of the Sphere

Keywords: Degrees (53), latitude (35), Earth (32), circumference (32), longitude (30)

> The people along the sand
> All turn and look one way;
> They turn their back on the land,
> They look at the sea all day.

Robert Frost
Neither Out Far, Nor In Deep

2.1 Introduction: Relative and absolute location

As is the case with Robert Frost's people along the sand, most of us are creatures of habit; we do not look out far, or in deep, at our real-world surroundings. We tend to categorize our surroundings in particular ways and to use those categorizations throughout our lifetimes as ways to reference position and navigation within our worlds. One way to think about our own mental structures is to consider the simple task of giving directions.

> Method 1: When you leave the campus of The University of Michigan, turn left on Washtenong, take it out of town past the bend in the road and beyond the split to the stadium. Keep going until you get to the new shopping mall then turn left onto Platt Parkway. Keep going on the parkway until you see a berm on the right and then enter the condo complex behind the berm. Turn left once inside. I live in the eighth condo from the end on the right side.
>
> Method 2: When you leave the campus of The University of Michigan, head southeast on Washtenong Avenue. Take Washtenong for 3.3 miles, southeast, and then due east, until you get to Platt Parkway. Turn north onto Platt Parkway and continue for 2.4 miles. Turn east onto Woodlawn Drive on the east side of the parkway. Head north on Woodlawn to 1452.

The first approach is a "relative" approach. The second is an "absolute" approach. Method 1 has the advantage of tying directions to landmarks that may be familiar. But, it is ambiguous. One set of directions can lead to many places. What is on the "left" when I am facing north, becomes on my "right" when I am facing south. Orientation is critical in this approach. In the "absolute" approach, however, orientation is irrelevant. Due east is the same, no matter which direction I am facing. One set of absolute directions leads to one location, and only to that one location. What is often a problem with an "absolute" approach is one of communication. Many people do not really think about the real world in terms of the cardinal points of the compass. Education is a great ally here and coupling it with contemporary technology offers ways to reinforce such education.

The examples given in "Method 1" and "Method 2" are extremes. Many would blend the two approaches, perhaps giving a street address as well as both landmarks and turn directions. Often, it is the case that folks who are not "good" with maps rely more on relative locational direction than do others. The ubiquity and ease of use of today's geotechnologies, with GPS-enabled maps on tablet computers, on smartphones, and in vehicles may be boosting the level of map literacy (although there are stories to the contrary). These devices rely on absolute location in order to offer clear and unambiguous direction to drivers. As with all useful devices, one still needs to think! Because the devices also offer turn-by-turn information, they actually may deter people from reading maps. Those who do not think about what they are doing may get lost or get into trouble on the road from blind, thoughtless, following of a "road routing recipe." As with any computerized tool, "use— not abuse!" Relative and absolute location may be already familiar in terms of giving simple local directions. The terms are, however, more far reaching than that. For example, absolute location can be given in hundreds of ways, depending on the coordinate system and datum used, both of which are discussed in various parts of this book.

Parallels and meridians were used in the previous chapter to describe the location of a point, *P*, on the Earth-sphere as being at the 3rd parallel north of the Equator and at the 4th meridian to the west of the Prime Meridian. Recall that while this strategy might serve to locate *P* according to one reference system, someone else might employ a reference system with a finer mesh. Ambiguity entered the picture, this time not in relation to orientation, but rather in relation to partition with regard to counting parallels and meridians. Many different descriptions, in how fine or coarse a mesh is created for the graticule, can lead to one location. To convert this system to an absolute system, which is replicable, we noted that we need to employ some commonly agreed-upon measurement strategy to standardize measurement—so that different ways of partitioning data lead to different answers. The human-constructed assumption that there are 360 degrees of angular measure in a circle permitted such standardization. The use of standard circular measure created a designation that was unique for *P*. One description leads to one location and only to one location.

These two examples, of local navigation around town and of global navigation on the Earth-sphere, may seem like scaled-up or scaled-down views of each other. In some ways they are. There is, however, a critical difference. In the local view, it was the concept of "orientation" that needed to be removed from the picture to have accurate directions. In the global view, it was the concept of "partition" that needed clarification. In both cases, moving from relative to absolute strategy solved the problem. Satisfactory solution was obtained in both cases when one description led to one location, as a "one-to-one" transformation.

2.2 Location and measurement: From antiquity to today

One well-known classical approach to mathematical geography appears in the work of Eratosthenes of Alexandria (c. 276–194 BC), a Greek mathematician, geographer, and astronomer. Eratosthenes made critical discoveries in both mathematics and geography that might seem, superficially at least, unrelated. His prime number sieve, or algorithm, permitted complete characterization of all whole numbers and offered, therefore, an understanding of the whole number system. He worked, as well, with spherical geometry, latitude and longitude, Earth–Sun relations, and a variety of geometric and trigonometric ideas. Eratosthenes sealed the gap between number theory and the real world with the idea of using abstract tools to understand more than what one could see, be that an infinity of whole numbers or an Earth larger than a hometown—his brain was wider than his sky—his East and West met in the abstract realm of pure mathematics. The transformational idea of understanding a whole, which one could never see, applies to his measurement of the circumference of the Earth (Arlinghaus and Arlinghaus, 2005).

Eratosthenes was the first known person to accurately measure the circumference of the Earth, doing so by measuring Sun angles at two different points along the same line of longitude (meridian) in North Africa. To perform his astonishingly accurate measurement of the circumference of the Earth, he used Euclidean geometry and simple measuring tools. Clearly, he understood in some way, the theory underlying coordinate systems and seasons expressed in **Chapter 1**. Spatial mathematics has deep roots in antiquity!

To understand how he might have had the intellectual combination of tools to achieve this remarkable feat, consider a bit of his background. He was born in Cyrene, now part of Libya, in northern Africa. After studying in Alexandria and Athens, he became the director of the Great Library in Alexandria in 236 BC. The library housed a great deal of the learned and compiled knowledge of the time. It was at the library that Eratosthenes read about a deep vertical well near Syene (now Aswan) in southern Egypt. Once a year at noon at this well, on the day of the summer solstice, the Sun lit the bottom of the well: The Sun was directly overhead, its rays shining straight into the well.

Eratosthenes then placed a vertical post (obelisk) at Alexandria, which was almost due north of Syene, and measured the angle of its shadow on the same date and time. Making the assumptions that (a) the Earth is a sphere and that (b) the Sun's rays are essentially parallel, Eratosthenes knew from the geometry of Euclid that the size of the measured angle (7°12′) equaled the size of the angle at the Earth's center between Syene and Alexandria. Also knowing that the arc of an angle of this size was approximately 1/50 of a circle, he then had to determine the distance between Syene and Alexandria. This was a difficult task during that time, due to different strides of camels and human error, and despite the best efforts of the King's surveyors, required years of effort. It was finally determined to be 5000 stadia. Then it became a straightforward matter to calculate the entire circumference: The angular measure between the two locations was 1/50th of a circle. Thus, Eratosthenes multiplied 5000 by 50 to find the Earth's circumference. His result, 250,000 stadia (about 46,250 km), was amazingly close to the accepted modern measurements (40,075 kilometers around the Equator and 40,008 kilometers around the poles).

Below, we show a more formal approach to the conceptual materials, and their implications, associated with the measurement that he is said to have made (different accounts give different details). See **Figure 2.1**.

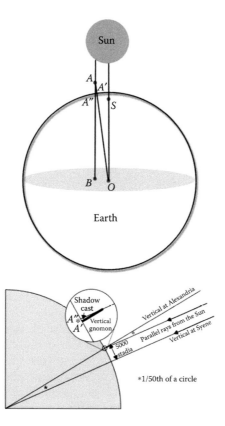

Figure 2.1 *Eratosthenes's measurement of the circumference of the Earth, based on a theorem of Euclid. In the top image, the Sun is shown as a small gray ball far away from the Earth (large ball). The Sun's rays are assumed parallel to each other as they strike the Earth's surface. The bottom image shows the detail of the configuration of obelisk (gnomon) and shadow in relation to the surface of the Earth.*

- Assume the Earth is a sphere.
  - The circumference of the sphere is measured along a great circle on the sphere.
  - Find the circumference of the Earth by finding the length of intercepted arc of a small central angle.
  - Find two places on the surface of the Earth that lie on the same meridian (or close to it): Meridians are halves of great circles.
  - Eratosthenes chose Alexandria and Syene, near contemporary Aswan (A' and S in **Figure 2.1**, respectively).
- Assume that the rays of the Sun are parallel to each other.
  - The Sun's rays are directly overhead, on the summer solstice (c. June 21), at 23.5 degrees north latitude.
  - Syene is located at about 23.5 degrees north latitude. Hence, on the summer solstice, sunlight will pass to the bottom of a narrow well (and it will not do so on other days), S.

- Alexandria is north of Syene. Thus, on June 21, objects at Alexandria will cast shadows whereas those at Syene will not.
- Eratosthenes focused on an obelisk (AA') or post located in an open area (**Figure 2.1**). The base of the post is at A' in Alexandria and its tip is at A (assuming the obelisk is a straight extension of OA'—that is, it is not a "leaning" tower). He measured the shadow ($A'A''$) that the obelisk cast, functioning in the manner of a gnomon on a sundial, and then measured the height of the obelisk (AA') (perhaps using a string anchored to the tip of the obelisk).
- According to Euclid, two parallel lines cut by a transversal have alternate interior angles that are equal. The Sun's rays are the parallel lines. One ray, at Alexandria, touches the tip of the obelisk and extends earthward toward the tip of the shadow of the obelisk, AA''. It is extended to AB in **Figure 2.1**. The other ray, SO, at Syene, goes into the well and extends abstractly to the center of the Earth, O. The obelisk, AA', also extends abstractly to the center of the Earth, O; thus, the line, AO, determined by the tip of the obelisk and the center of the Earth is a transversal cutting the two parallel rays, SO and AB, of the Sun.
- Angles (BAO) and (SOA) are thus alternate interior angles in the geometric configuration described above; therefore, they are equal.
- Trigonometry permits angle measurement (see review in subsequent section). Use the length of the obelisk shadow and the height of the obelisk to determine angle (BAO); triangle ($A'AA''$) is a right triangle with the right angle at A'. Thus, tan ($A'AA''$) = (length of shadow, $A'A''$)/(height of obelisk, AA'). Eratosthenes's measurements of these values led him to conclude that the measure of angle ($A'AA''$) was 7 degrees and 12 minutes and so therefore it was also of angle (BAO).
  - The value of 7 degrees and 12 minutes is approximately 1/50th of the degree measure of a circle. Since he assumed that Alexandria and Syene both lay on a meridian (half of a great circle), it followed that the distance between these two locations was 1/50th of the circumference of the Earth.
  - Eratosthenes calculated the distance between Alexandria and Syene using records involving camel caravans. The distance he used was 5000 stadia. Thus, the circumference of the Earth is 250,000 stadia, which translates to somewhat more (depending on how ancient units convert to modern units) than current accepted values although it is remarkably close.

Many of the assumptions made by Eratosthenes were not accurate; apparently, however, underfit and overfit of error balanced out to produce a good result. For example, Syene and Alexandria are not on the same meridian, and Syene is not at exactly 23.5 degrees north latitude. To gain a deeper understanding of the principles involved in this measurement, you have the opportunity to participate in the activity or "practice" in the following section. The activity involves using a GPS receiver to measure the distance between two lines of latitude, determining the distance between the two, and calculating the circumference of the Earth from those two readings. Thanks to the accuracy of GPS technology, you do not have to walk nearly as far as Eratosthenes' surveyors did—only a few hundred meters, at most. After determining the polar circumference, you can calculate the equatorial circumference, and then the mass and volume of the Earth. From your stroll with an ordinary GPS receiver and your calculations, your circumference estimate should be within 1% of the accepted value of the circumference of the Earth, providing a perfect example of the power of combining geography and mathematics.

2.3 Practice: Measuring the circumference of the Earth using GPS

With a GPS receiver, you and a group of colleagues can incorporate some of Eratosthenes' methods to accurately measure the circumference of the Earth to within 1% of the accepted value. This activity involves going into the field, to a broad open area, to calculate the circumference in a variety of different ways. Emulate Eratosthenes' methods in the field using modern technology, specifically GPS. By so doing, you are integrating geography, mathematics, Earth science, and physics to solve an applied problem.

We will calculate and consider both the polar and equatorial circumferences of the Earth as well as the Earth's mass and volume. At the end of the field work, evaluate how close your measurements are to the accepted values of the Earth's circumference, mass, and volume!

2.3.1 Measuring the Earth's polar circumference using Table 2.1

1. Participants should organize into teams of two.
2. In each team, you will mark waypoints using the UTM (Universal Transverse Mercator) coordinate system, where the units are in meters. The numbers for UTM represent eastings (relative to the Central Meridian in the UTM zone), and northings (relative to the Equator). Important: Set each unit to the datum WGS 84 so that all participants are working with the same datum (model of the Earth's shape). To obtain decimal degree values (latitude–longitude coordinates) for each point, you can set the units to decimal degrees (DD) once you are back in the laboratory and access the waypoint information page for that waypoint to obtain the coordinates in decimal degrees. Alternatively, you could simply change the GPS units in the field and gather the waypoints in decimal degrees. Or, you could organize in teams of two with two GPS's, one set to latitude–longitude decimal degrees, and the other in UTM eastings and northings.
3. Locate a safe place away from buildings, cliffs, trees, power lines, and other obstacles such that you will be able to walk in a north/south direction for at least 200 meters—the longer the better. You can use

Table 2.1 Measuring the Earth's Polar Circumference

North point (latitude: *decimal degrees*)	a.
North point (longitude: *decimal degrees*)	b.
North point (UTM northing: *meters*)	c.
North point (UTM easting: *meters*)	d.
South point (latitude: *decimal degrees*)	e.
South point (longitude: *decimal degrees*)	f.
South point (UTM northing: *meters*)	g.
South point (UTM easting: *meters*)	h.
Distance in decimal degrees [a–e°]	i.
Distance in meters [c–g *meters*]	j.

Knowing the ratio

$$\frac{\text{Distance in decimal degrees } [i]}{360°} = \frac{\text{Distance in meter } [j]}{[PC]\text{meters},}$$

we can solve to find the Earth's polar circumference

$$[PC]\,km = \frac{([j]\text{meters} * 360°)}{[i]°} * \frac{1}{1000} \qquad \text{k.}$$

To calculate % error (assuming a true polar circumference of 40,008 km)

$$[PC]\%\,error = \frac{|\,[k]\,km - 40,008\,km\,|}{40,008\,km} * 100 \qquad \text{l.}$$

Note: Do not round off! Please retain a minimum of five places to the right of the decimal point.

the compass on a GPS set to true north or monitor the GPS coordinates to make sure that you are following a true north–south track. If you are following a true north–south track, the eastings should not change as you are walking.

4. Take a GPS reading at the north end of your chosen track (Note: If you can "average" the points and then mark the waypoint, please do so (and make sure that WAAS (Wide Area Augmentation System) correction is enabled)). Name the waypoint with a meaningful name so that you can identify it later. Record the UTM coordinates in **Table 2.1**. You can record the latitude and longitude from this stored waypoint later in the laboratory or you can switch units in the field and gather a new waypoint in decimal degrees (see step 2 above). Enter the value for decimal degrees in **Table 2.1**.
5. Repeat for the south end of your chosen track.

2.3.2 Measuring the Earth's equatorial circumference using Table 2.2

1. Follow steps 1 and 2 for "Measuring the Earth's Polar Circumference" above.

Table 2.2 Measuring the Earth's Equatorial Circumference

West point (latitude: *decimal degrees*)	a.
West point (longitude: *decimal degrees*)	b.
West point (UTM northing: *meters*)	c.
West point (UTM easting: *meters*)	d.
East point (latitude: *decimal degrees*)	e.
East point (longitude: *decimal degrees*)	f.
East point (UTM northing: *meters*)	g.
East point (UTM northing: *meters*)	h.
Distance in decimal degrees [b–f°]	i.
Distance in meters [h–d *meters*]	j.
Knowing the ratio	k.

$$\frac{\text{Distance in decimal degrees [i]}}{360°} = \frac{\text{Distance in meters [j]}}{\text{[LC] meters}}$$

We can solve for the circumference of the Earth *along this line of latitude* [LC] in *kilometers*

$$[LC]\,km = \frac{([j]\,meters * 360°)}{[i]°} * \frac{1}{1000}$$

and to determine the equatorial circumference [EC] in *kilometers (use degrees not radians)*

$$[EC]\,km = \frac{[LC]}{\cos(\text{latitude [a]})}$$

To calculate % error (assuming a true equatorial circumference of 40,075 km)   l.

$$[EC]\,\%\,error = \frac{|\,[k]\,km - 40{,}075\,km\,|}{40{,}075\,km} * 100$$

Note: Remember your work on the activities in **Chapter 1**: Do not round off! Retain a minimum of five digits to the right of the decimal point.

---

2. Locate a safe place away from buildings, cliffs, trees, power lines, and other obstacles such that you will be able to walk in an east/west direction for at least 200 meters—the longer the better (Note: You may want to start at one of the points used for the polar circumference calculation).

   You can use the compass on a GPS set to true north or monitor the GPS coordinates to make sure that you are following a true east–west track. If you are following a true east–west track, the northings should not change as you are walking.

3. Take a GPS reading at the east end of your chosen track (Note: If you can "average" the points and then mark the waypoint, please do so (and make sure that WAAS correction is enabled)). Name the waypoint with a meaningful name so that you can identify it later. Record the UTM coordinates in **Table 2.2**. You can record the latitude and

longitude from this stored waypoint later in the laboratory or you can switch units in the field and gather a new waypoint in decimal degrees (see step 2 above). Enter the value for decimal degrees in **Table 2.2**.

4. Repeat for the west end of your chosen track.

2.3.3 For further consideration: Polar circumference and equatorial circumference

1. How does your calculated polar circumference compare with the accepted polar circumference? You should be within 1% of the accepted value of 40,008 kilometers (24,860 miles). What are some reasons for why your answer may not be exactly the same as the accepted circumference of the Earth? These include the fact that GPS is not a perfect system, and while performing the activity multiple times and averaging the values that you obtained will minimize error, it will not eliminate error. GPS signals may bounce off buildings before reaching your receiver (known as multipath error), heavy tree cover may interfere with the GPS signals, or there may be fewer than the optimal number of GPS satellites in view during the time of your experiment.

2. How does the accepted polar circumference compare with Eratosthenes' calculated circumference? Eratosthenes calculation of 46,250 kilometers is 15.6% higher than the accepted modern value. The reasons why Eratosthenes' circumference differs from the accepted polar circumference include the reliance on the imperfect methods of counting footfalls from people and camels from the surveyor team that trekked across the desert to determine the distance between the two cities, the fact that Alexandria is not due north of Syene, and the possibility that he may have thought that the Earth was perfectly spherical and thus did not reduce the value of the equatorial circumference to determine the polar circumference.

3. What do your polar and equatorial circumference calculations tell you about the shape of Earth? Do some research and discuss why the Earth has this shape. Your calculations alone should indicate that the Earth is not a perfect sphere, but rather, that the Earth bulges around its middle. Your research will show that the Earth is an oblate spheroid, and the larger equatorial circumference has to do with the rapid rotation of the Earth and its molten core. The equatorial bulge is even more pronounced with the gaseous planets Jupiter, Saturn, Uranus, and Neptune, and with stars like our own Sun.

4. Using your calculated values, how long would it take to walk around the Earth along the polar circumference (in days)? You can assume that you can walk an average of 5 km/h, or if you want to check your actual average walking speed, use the speed function in your GPS while you are

walking to find out what your average walking speed is. To determine this, set the receiver to the page where you can determine how fast you are moving (km/h). If you could sustain 5 km/h, each day you could cover 120 km. You could thus walk around the Earth in 40008/120, or 333.4 days. However, traversing the ice at the poles, mountain ranges, and the oceans would significantly slow your progress!

5. How long would it take to walk along the equatorial circumference (in days)? At the same rate of speed that was used above, you could walk around the Equator in 385.4 days. Here, mountains in Ecuador, Kenya, and Indonesia would slow your progress, and once again you would have that pesky problem of traversing the oceans.

6. How long would it take for you to walk along the circumference of your line of latitude in **Table 2.2** (in days)? At 40 degrees north latitude, the circumference is 30,656.92 km. Therefore, it would require 255.5 days at 120 km/day.

2.3.4 Determining the mass and volume of the Earth using Table 2.3

With a GPS, you can also determine the mass and volume of the Earth. Follow steps a through f in **Table 2.3**.

2.4 Measuring positions on the Earth surface, and fractions

Most of us studied fractions from primary school to university level and use them in our everyday lives. However, when engaged in field work, it is important to feel comfortable with such matters so that one can devote full attention to what is going on in the field and need not have to stop and consider fractional conversions and transformations. Thus, we offer a quick review of fractions emphasizing conversion from degree/minute format to decimal degrees and the addition of fractions with different denominators. We illustrate the concepts through worked examples.

For example, consider the conversion of 42 degrees, 31 minutes, 47 seconds, to decimal degrees. This entity can be written in fractional form as $42 + 31/60 + 47/3600$. Use a calculator to evaluate the answer. Where did the value of 3600 come from? Consider the standard clock. It is in base 60, with 60 seconds creating one minute, and 60 minutes creating one hour. Similarly, measurements on the Earth are in degrees, minutes, and seconds (DMS), also in base 60, but with degrees, minutes, and seconds measured as distances instead of time units. So, here, 3600 is 60 squared, required because 60 seconds create one minute, and 60 minutes create one degree. So, the answer is simply (approximately) $42 + 0.5167 + 0.013056 = 42.529756$.

Now, reverse the process: Convert 42.53 to degrees, minutes, and seconds. Seconds is the finest unit. There are 3600 seconds in a degree because there are

Table 2.3 Determining the Mass and Volume of the Earth

Determine the length of the polar circumference	a.
Knowing that the circumference of a circle is	b.

$$2 * \pi * R$$

where *R* is the radius of the Earth

$$[R_{km}] \, km = \frac{[a] \, km}{(2 * \pi)}$$

$$[R_m] \, meters = [R_{km}] * 1000$$

Mass   c.

Knowing that mass is

$$Mass = \frac{(acceleration * R^2)}{G}$$

where acceleration due to gravity is 9.8 m/s², and *G*, the constant of proportionality, is 6.67×10^{-11} m³/kg s²

You can determine the mass (kg) of the Earth using the Earth's radius $[R_m]$ from [b].

To calculate the mass % error (assuming a true mass of 5.98×10^{24} kg)   d.

$$\% \, error = \frac{|[c] \, kg - 5.98 \times 10^{24} \, kg|}{5.98 \times 10^{24} \, kg} * 100$$

Volume   e.

Knowing that volume is

$$Volume = 4/3 * \pi * R^3$$

You can determine the volume (km³) of the Earth using the Earth's radius $[R_{km}]$ from [b].

To calculate the volume % error (assuming a true volume of 1.0975095×10^{12} km³)   f.

$$\% \, error = \frac{|[e] \, km^3 - 1.0975095 \times 10^{12} \, km^3|}{1.0975095 \times 10^{12} \, km^3} * 100$$

---

60 minutes in a degree and 60 seconds in a minute. Consider the value 42.53 as equal to $42 + 5/10 + 3/100$. Convert the denominators of the fractions to 60 and to 3600. Do so by multiplying each proper fraction by the number 1, written in one case as 6/6 and in the other case as 36/36. In the case of the minutes, 5/10 is the same as 30/60. In the case of the seconds, 3/100 is the same as 108/3600. Thus, $42.53 = 42 + 30/60 + 108/3600$ and this value translates to 42 degrees, 30 minutes, 108 seconds. But, 108 seconds reduces to one minute and 48 seconds. So, the answer is 42 degrees, 31 minutes, 48 seconds. The discrepancy in value in reversing the process is due to round off. The method is accurate.

In the last paragraph, we multiplied fractions by 1, written one time as 6/6 and another time as 36/36. Creative use of multiplying by the number 1 is the root of addition of fractions. Often, unfortunately, one learns to add fractions before learning to multiply them (perhaps as a simple curricular extension of how one learns to work with integer arithmetic). That pattern is unfortunate because multiplying fractions is straightforward and natural; further, it is critical to know how to multiply fractions in order to add fractions, systematically, with different denominators. To add 1/3 and 1/4 multiply the first fraction by 4/4 and the second one by 3/3 yielding $4/12 + 3/12 = 7/12$. For those wishing to think of fractions in terms of pieces of pie, cut one pie into thirds and split each third into quarters so that the entire pie is split into 12 pieces. Split another pie into quarters and split each quarter into thirds so that the entire pie is split into 12 pieces. Now the two pies are commensurable, each having 12 pieces of equal size. From the first, take four pieces and from the second take three pieces; so the sum is $4 + 3$ or 7 of the pieces that are each 1/12th of the pie. Again, the answer is 7/12. The idea of using the number 1 expressed in creative formats is a powerful numerical transformation. Look for it throughout mathematics from the simple to the advanced.

2.5 Other common coordinate systems

We have discussed two different ways of representing latitude and longitude: As degrees, minutes, and seconds, and as decimal degrees. As we noted above, these are commonly abbreviated as DMS and DD, respectively. A third common way of representing latitude and longitude is called decimal minutes (DM). As the name implies, this method uses degrees, minutes, and then fractions of a minute as its notation, such as 105 degrees 30.5 minutes. Recall our work earlier in this chapter using the value of 42 degrees, 31 minutes, 47 seconds in DMS notation. In DD format, it is represented as 42.529756. In DM, this can be written in fractional form as 42 degrees, 31 + 47/60, or 42 degrees, 31.7833 minutes. DM format is used by specific groups such as the geocaching community to locate hidden treasures with GPS technology, and it is a common format that most GPS receivers can display.

Five other coordinate systems are worth discussing here because they are other manifestations of absolute location. Any coordinate system requires three things to represent absolute location: An origin point, two sets of axes, and a unit of measurement. Latitude and longitude are referenced from the origin point of (0, 0), and its two sets of axes are the Equator (the X-axis) and the Prime Meridian (the Y-axis). Their unit of measurements can be in any of three formats—degrees minutes seconds, decimal degree, and decimal minutes.

At the beginning of this chapter, street addressing was discussed. Street addressing in many places around the world is a manifestation of a coordinate system. In most cities, an east–west street or an imaginary line divides the north from the south sectors of the city, and a north–south street or imaginary line divides the east from the west sectors. The place where these lines intersect does not have to be in the center of the city, and oftentimes is not, due to the historical pattern of the way the city grew after the coordinate system was devised, or because the city founders wanted all or most of the addressing for the central part of the city to be on one side of one of the axes, or because of a natural feature such as a river, or for other reasons. In many cities, addresses are referenced from these two axes, typically with a range of 100 between each block. Often, even and odd were given designated sides; for example, in Denver, Colorado, addresses on the east and south sides of streets are even, while addresses on the north and west sides of streets are odd. Thus, a house on Broadway that is three blocks north of the X-axis on the east side of the street could have an address of 308 North Broadway.

It is worth noting that unlike latitude and longitude, the addressing system sometimes is not rigorously followed in a city that has officially adopted it, and may be totally ignored in other cities. Indeed, in some places, addresses do not follow a regular system, and in others, such as Tokyo, houses are given successive numbers depending on when they were built on the block. Hence, the system there is historical rather than geographical. In scores of other cities and rural areas around the world, there are no street addresses at all, and in many cities, very few of the streets even have names. In the United States, until relatively recently, most rural addresses used the "route and box" system. The route represented the route of the postal carrier, and the box was a physical mailbox along that route; for example, Route 4, Box 358, Mequon, Wisconsin. While the boxes and the routes may have been numbered consecutively, there was much ambiguity and duplication; it was physically difficult to locate a house or apartment with this system. When there is an addressing system in place, it can serve as an absolute and unique identifier of residences, industrial sites, and businesses in that location. Addressing is so valued as a way of locating homes in the event of an emergency that many rural addressing programs over the past 20 years in the United States have been undertaken. These addressing programs are aimed at converting the "route and box" addresses to "house number, street" addresses.

A second coordinate system is the State Plane Coordinate System. Used exclusively in the USA, this system consists of a set of 124 separate coordinate systems, or "zones," and each state is covered by one or more of them. The boundaries of the zones usually follow county lines. A Cartesian coordinate system is created for each zone by establishing an origin some distance (usually 2,000,000 feet) to the west of the zone's central meridian and some distance to the south of the zone's southernmost point. This ensures that all coordinates within the zone will have positive values. The X-axis running through this origin runs east–west, and the Y-axis runs north–south. X distances are typically called eastings (because they measure distances east of the origin) and Y distances are typically called northings (because they measure distances north of the origin). Some zones use feet, and others use meters. The system is employed widely for spatial data used by and distributed by state government agencies, in part because it uses a simple Cartesian coordinate system to specify locations and in part because it is highly accurate. A typical coordinate might be 310,200 feet north, 88,410 feet west, Wisconsin Central Zone, NAD 83 HARN (indicating the horizontal datum used), US Survey feet (indicating the units). In a similar fashion, many other countries have their own reference system, such as the British National Grid or the New Zealand Map Grid, for use in topographic mapping and digital spatial data.

A third additional coordinate system is the Universal Transverse Mercator (UTM) system. The US National Geospatial Intelligence Agency adopted this grid for military use throughout the world, dividing it into 60 north–south zones. Each zone covers a strip 6° wide in longitude so that all 360° of the Earth could be completed. The zones are numbered consecutively beginning with Zone 1, between 180° and 174° west longitude, progressing eastward to Zone 60, between 174° and 180° east longitude. The conterminous US States are covered by 10 zones, from Zone 10 on the west coast through Zone 19 on the east coast. In each UTM zone, coordinates are measured east and north in meters as "eastings" and "northings." Let us take the second value first, the northing, because it is more straightforward. This northing value is measured continuously from zero at the Equator, in a northerly direction. Thus, a northing of 3,446,250 is exactly 3,446,250 meters, or 3,446.25 kilometers, north of the Equator. To avoid negative numbers for locations south of the Equator, the Equator is assigned an arbitrary false northing value of 10,000,000 meters. Now let us consider the first value, the easting: A central meridian through the center of each 6° zone is assigned an easting value of 500,000 meters. Why 500,000? Again, it is simply given a high enough value to ensure that no easting coordinate will have a negative number. Values to the west of this central meridian are less than 500,000; to the east, they are more than 500,000. An easting value of 529,781 is 29,781 meters, or 29.781 kilometers east of the central meridian in that zone. Thus, a full UTM coordinate example is 391,000 meters easting, 3,456,789 meters northing, UTM Zone 13. The zone is needed because the same coordinates (such as 391,000 meters easting, 3,456,789 meters northing) can exist in more than one zone (such as Zone 13 and Zone 14). In **Figure 2.2**, Edmonton, Alberta, Canada is designated as a point and three different descriptions for it are given: Approximate address, latitude/longitude, and UTM coordinates.

The fourth coordinate system that is in wide use at least in the United States is the Public Land Survey System (PLSS). It began in eastern Ohio with the Land Ordinance of 1785 with the opening up of the American West, and was the primary system used to subdivide the land for sale to homesteaders over the next century and a half. It continues to be used to the present day. It covers about 2/3 of the country, primarily in the central and western portion, with the exception of Texas. It is referenced off of a series of meridians and baselines, with the land divided up into ranges and townships of 36 square miles, and sections of one square mile. Each section

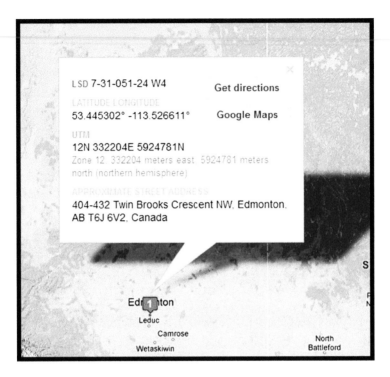

Figure 2.2 *Edmonton, Alberta, Canada. Description using different coordinate systems. Source: BaseLoc (www.baseloc.com), division of GPS Police (www.gpspolice. com), Calgary, AB.*

is then divided into quarter-sections, and each quarter section into further quarters and halves. The resulting coordinate for a parcel of land could be "Northwest 1/4 of the southeast 1/4, Section 6, Township 2 South, Range 55 West, Sixth Principal Meridian." Therefore, this system, while useful for identifying tracts of land on property deeds, is restricted to land areas only. The land areas can be quite small and specific, but the system is not meant to identify individual points.

The Military Grid Reference System (MGRS) is the fifth and final system to be treated here, used by North Atlantic Treaty Organization (NATO) militaries for determining absolute location. MGRS is derived from the UTM system but covers the entire Earth and uses a different labeling convention than does UTM. It consists of three parts: A grid zone designator, a 100,000 square meter identifier, and a number that corresponds to the easting and northing in that cell. The resolution of a full MGRS coordinate is a one meter square, and if digits are left off, it can be used to represent blocks of 10, 100, 1000, or larger squares. An example of a one meter MGRS address is 4QFJ1234567890, where 4Q is the grid zone designator, FJ is the 100,000 square meter identifier, and the other numbers allow a one square meter cell to be pinpointed.

In the following activity section, you will have the opportunity to use different coordinate systems, such as latitude and longitude, the Military Grid Reference System, and the Public Land Survey System.

2.6 Practice: Coordinates using different systems

Access the National Map Viewer from the USGS (http://viewer.nationalmap.gov/viewer). Click on the Find Coordinates tool and click on St Louis. What are the coordinates of St. Louis in Decimal Degree Latitude–Longitude (DD), in Degrees Minutes Seconds Latitude–Longitude (DMS), and in the Military Grid Reference System (MGRS)? These readings are, depending on where in the city you are clicking, 38.62484 degrees north, 90.20135 degrees west in DD, and 38 degrees 37 minutes 29.427 seconds north, 90 degrees 12 minutes 4.877 seconds west in DMS. It is 15SYC4363578862 in MGRS. Calculate the DM coordinates yourself using the values you just discovered using the National Map Viewer. The DM coordinates are calculated as follows: 29.427/60 = .49045, and therefore the latitude of St. Louis in DM is 38 degrees and 37.49045 minutes north. The longitude is 4.877/60 or .081283, so the DM longitude of St. Louis is 90 degrees 12.081283 minutes west. What are the coordinates of your own community according to each of these four systems?

Find coordinates using several different methods by using GPS Visualizer (http://www.gpsvisualizer.com/calculators). This tool allows coordinates to be calculated and mapped in several different systems, as well as distances between two points on the globe.

Use the Find Latitude and Longitude (http://www.findlatitudeandlongitude.com/), and search for other latitude and longitude tools. How do these tools compare? Which tools give insight into the algorithms used? If they do not give information about the algorithms used, can you trust them? Keep in mind one of our central themes of this book: Know your data, and know the underlying methods and assumptions that led to the creation of the data in the first place. Web-based tools are rich and varied, but need to be carefully scrutinized.

Examine the Principal Meridians and Baselines map at the Bureau of Land Management on http://www.blm.gov/wo/st/en/prog/more/cadastralsurvey/meridians.html. This map shows the meridians and baselines that make up specific zones in the Public Land Survey System (PLSS) in specific states. What principal meridian and baseline covers most of the land of Wyoming, Colorado, Kansas, and Nebraska? This is the Sixth Principal Meridian and Baseline. How many zones cover the state of California? The map shows three of them: The Mount Diablo, covering about 2/3 of the state, the Humboldt covering the northwest corner, and the San Bernardino, covering the southern third.

Use ArcGIS Online (http://www.arcgis.com/home), start a new map, use "Add" and search for the "USA Topo" to find the USGS topographic maps. Add the USA Topo map layer. Then search for the location "Hiawatha, KS." Zoom in until you can see the PLSS system with its sections, townships, and ranges identified. Indicate the PLSS section for the location of the courthouse in Hiawatha. The map shows the courthouse and most of the town as occupying Section 29. For obtaining the township and range for Hiawatha, you may need to consult additional map sources. Or, search and add the layer "Kansas Public Land Survey System" to your map. Can you determine the township and range for Hiawatha? Hiawatha is in Township 2 South Range 17 East.

2.7 Theory: Visual trigonometry review

Another part of mathematics that most of us learned in high school and used in college in calculus courses and elsewhere involves trigonometric functions. They were important in the time of Eratosthenes and permitted him to come to his remarkable estimate of the circumference of the Earth. They are as important today as they were then. They serve as a theoretical backbone of much of the capability of Earth measurement that has enabled wonderful technological advances such as handheld Global Positioning devices that let us find our way around the complex world of today. Thus we offer a straight-forward visual display of trigonometry to ensure that all are working from the same conceptual base. A visual approach can make things clear that might otherwise seem mysterious.

The geometric arrangement in **Figure 2.3a** shows the basic set of entities (based on a unit circle in the Euclidean plane) that we shall use to illustrate the geometric origins of trigonometric functions (Arlinghaus and Arlinghaus, 2005). Notice the importance of the idea of "axis" and "co-axis." Here, the word prefix "co" stands for "complementary," as in "at a right angle." The significance of the naming will play out in the naming of trigonometric functions, such as "sine" and "co-sine" (cosine). As with datums in **Chapter 1** and here, careful attention to axis arrangement is important.

Figure 2.3b shows the geometric derivations of the trigonometric functions for the angle θ. The length of the green line, dropping from P to the axis, measures the sine of θ: It is the opposite side of a right triangle over the hypotenuse (here, a radius of the unit circle). Draw a tangent line to the circle at (1,0) on the axis. The length of the red line, from the axis to P along the line tangent to the circle, measures the tangent of θ. The length of the blue line, from the origin to the red line (measured along the secant line), measures the secant of θ. Using the Pythagorean Theorem, the reader should verify that it follows easily from the geometry in **Figure 2.3b** that $\sec^2 \theta - \tan^2 \theta = 1$.

Figure 2.3c shows the geometric derivations of the trigonometric co-functions (where "co" stands for "complementary"). The length of the green line, dropping from P to the co-axis, measures the sine of co-θ; hence, cosine of θ. Draw

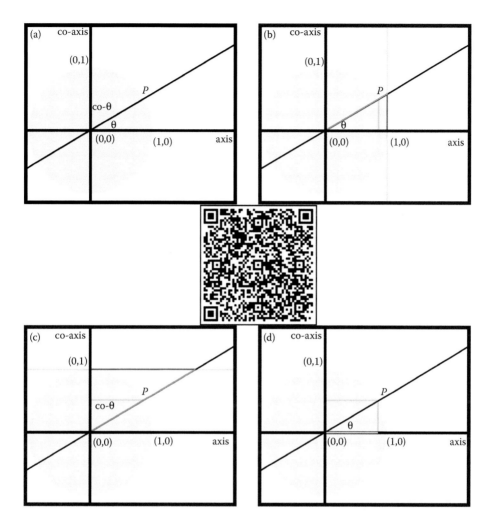

Figure 2.3 *(a) Unit circle, axis, and complementary (orthogonal) axis designated as co-axis. A secant line intersects the circle at P and forms an angle of* θ *with the axis and an angle of co-*θ *with the co-axis. (b) Derivations of sine (green), tangent (red), and secant (blue) functions of angle* θ*. (c) Derivations of co-sine (green), co-tangent (red), and co-secant (blue) functions of co-angle* θ*. (d) Shows right triangle interpretation of cosine as adjacent side over hypotenuse. Source: Modified from Arlinghaus, S. L. and W. C. Arlinghaus, 2005. Spatial Synthesis: Centrality and Hierarchy. Volume I, Book 1. http://www.imagenet.org, Introduction. QR code links to animation from that book, http://www-personal.umich.edu/~copyrght/image/books/Spatial%20Synthesis/trig/anisandytrig.gif*

a tangent line to the circle at (0,1) on the co-axis. The length of the red line, from the co-axis to P along the line tangent to the circle, measures the tangent of co-θ; hence cotangent of θ. The length of the blue line, from the origin to the red line (measured along the secant line), measures the secant of co-θ; hence cosecant of θ. Using the Pythagorean Theorem, the reader should verify that it follows easily from the geometry in **Figure 2.3c** that $\csc^2 \theta - \cot^2 \theta = 1$.

Note that the green line in **Figure 2.3c** fits exactly into the space in **Figure 2.3b** measured along the axis from the origin to the green line (**Figure 2.3d**)—that is, that $\cos \theta$ may be viewed as the adjacent side of a right triangle over the hypotenuse (here, a radius of the unit circle). Now, the reader can verify in a straightforward visual manner that $\sin^2 \theta + \cos^2 \theta = 1$ (**Figure 2.3d**). These three trigonometric identities serve as a basic set from which others may be verified by reducing them, through algebraic manipulation, to one of these three forms.

In the next section, you will have the opportunity to determine the length of one degree on the Earth.

2.8 Practice: Find the length of one degree on the Earth-sphere

- One degree of latitude, measured along a meridian or half of a great circle, equals approximately 69 miles (111 km). Therefore, one minute is 69/60, or just over a mile (1.15 miles), and one second is around 100 feet ((1.15 × 5280 = 6072 feet, divided by 60, or 101.2 feet), a pretty precise location on a globe with a circumference of 25,000 miles). Calculation: 25,000/360 = 69.444 miles in one degree of latitude along a meridian. This value varies slightly because the Earth is not actually a sphere; hence, the use of the generalized value of 25,000 for the Earth's circumference.
- Because meridians converge at the poles, the length of a degree of longitude varies, from about 69 miles at the Equator to zero at the poles (longitude becomes a point at the poles). Calculation: At latitude θ, find the radius, r, of the parallel, small circle, at that latitude. The radius, R, of the Earth-sphere is $R = 25,000/(2*\pi) = 3978.8769$ miles (assuming the circumference of the Earth-sphere at 25,000 miles). Thus, $\cos \theta = r/R$ (using a theorem of Euclid that alternate interior angles of parallel lines cut by a transversal are equal). Therefore, $r = R \cos \theta$. Then, the circumference of the small circle is $2r*\pi$ and the length of one degree at θ degrees of latitude is $2r*\pi/360$.

    Can you determine the length of one degree of longitude along any latitude line?
    - For example, at 42 degrees of latitude, $r = R \cos 42 = 3978.8769 * \cos 42 = 2956.882$. Thus, the circumference of the parallel at 42 degrees north latitude is approximately $2\pi*2956.882 = 18578.6205$ miles. Therefore, the length of one degree of longitude, measured along

the small circle at 42 degrees of latitude, is: 18578.6205/360 = 51.607 miles.

- This particular calculation scheme is a rich source of problems and puzzles using geometry and trigonometry. Consider the following question: At what latitude, angle θ, is the length of one degree of longitude exactly half the value of one degree of longitude at the Equator (69/2 = 34.5 miles)? These general steps are suggested as guides for the reader: $34.5 = 2\pi r/360$ so that $r = 1976.7$, the radius of the parallel at the desired latitude. Thus, $r = R\cos\theta$ becomes $\cos\theta = 1976.7/3978.8769 = 0.496798$, rounding to 0.5. Hence, $\theta = 60$ degrees.

In the next section, you will have the opportunity to further explore the connections between geography and mathematics through the exploration of seasons, and specifically, calculating Sun angles for different latitudes and at different times of the year.

2.9 Practice: Determine Sun angles at different seasons of the year

- The position of the Sun in the sky. On the date of the northern hemisphere's summer solstice, usually on or close to June 21, the direct ray of the Sun is overhead, or perpendicular to a plane tangent to the Earth-sphere, at 23.5 degrees north latitude. The angle of the Sun in the sky at noon with the ground at that latitude on that day is 90 degrees. What is the angle of the Sun in the sky, at noon on June 21, at 42 degrees north latitude?
  - Again, simple geometry and trigonometry solve the problem for this value and for any other. Use the fact that 42 – 23.5 = 18.5 degrees; that there are 180 degrees in a triangle (look for a right triangle with the right angle at 42 degrees north latitude); and, that corresponding angles of parallel lines cut by a transversal are equal. The answer works out to be 90 – 18.5, or 71.5 degrees. Thus, on June 21 at local noon, in the northern hemisphere at 42 degrees north latitude the Sun will appear in the south at 71.5 degrees above the horizon. In the southern hemisphere at 42 degrees south on this day it will appear in the northern sky at 42 + 23.5 = 65.5, then 90 – 65.5 = 24.5 degrees above the horizon.
  - Between the tropics, closer to the Equator, some interesting situations prevail (Arlinghaus, 1990). The use of this technique can be important in calculating shadow and related matters in electronic mapping including in, but not limited to, the formulation of virtual reality models of tall buildings.
- Further Directions:
  - The north and south poles are the Earth's geographic poles, located at each end of its axis of rotation. All meridians meet at these poles. The magnetized compass needle points to either of the Earth's two

magnetic poles. The north magnetic pole is located in the Queen Elizabeth Islands group, in the Canadian Northwest Territories. The south magnetic pole lies near the edge of the continent of Antarctica, off the Adélie Coast. The magnetic poles are constantly moving. What are the implications of this fact for the stability of our graticule?

In the following section, you will have the opportunity to work with measurement, the graticule, and with map projections using a variety of web-based GIS tools.

2.10 Practice: Work with measurement, the graticule, and map projections

In this section, you will have the opportunity to measure units off of the graticule using a web-based GIS. Open a web browser and access www.arcgis.com (ArcGIS Online). In the search box, enter "World Relief Map for Schools" and open it. Zoom out until you see the lines of latitude and the lines of longitude represented in a spacing of 15 degrees apart. On a map in the Mercator projection, such as this, notice that the lines of latitude and longitude are represented as straight lines. We will discuss map projections in detail in a later chapter, but notice how the lines of longitude are spaced evenly while the lines of latitude get farther apart as one approaches the North Pole and the South Pole. What do you notice about the sizes and shapes of land masses near the North Pole and the South Pole? Why do you suppose these distortions occur? The sizes and shapes become distorted near the poles because of the fact that any map has to distort the bounded curved surface of the Earth onto an unbounded flat paper or digital map.

Verify the numeric value of the latitude and longitude lines by clicking on each of them. Select the tool that shows the individual map layers that make up your map, and expand the graticule layer several times to see that this layer is actually composed of several layers, as shown in **Figure 2.4**.

These layers are scale dependent. As you zoom in, a finer, higher resolution mesh of latitude and longitude lines appears. Zoom to the Equator in South America by holding the Shift key and simultaneously dragging a box with the mouse. Or, search for Quito, Ecuador, and find the Equator nearby. What is the finest mesh of latitude and longitude lines that appears? According to this map, it is a one degree mesh. Consider the discussion of latitude and longitude reference lines earlier in this chapter. Is the latitude–longitude grid you are now examining the finest mesh that could be generated? Why or why not? There are an infinite number of finer graticules that could be generated; for example, a half-degree, 10 minute, or even one second mesh.

Use the measure tool to find the distance between one degree of longitude and the next degree of longitude to the east or west. Indicate the units you have

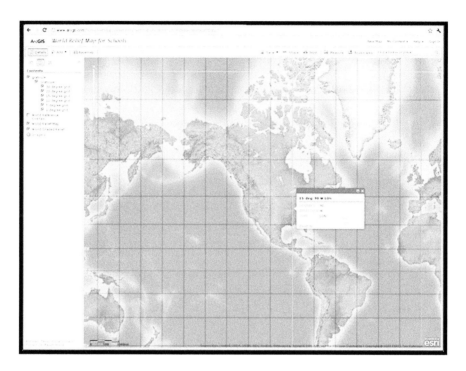

Figure 2.4 *Map in ArcGIS Online showing uneven spacing between lines of latitude, but even spacing between lines of longitude, on this modified Mercator projection. Source of base maps: Esri software.*

chosen (miles or kilometers). How closely does this measured distance match the distance indicated in the discussion earlier in this chapter? For example, at Quito, the distance appears to be around 110 km, and the accepted value we discussed earlier was 111.319 km. What are a few reasons why they do not match? It depends on where you click with your mouse, which is in itself an imperfect science, and the map projection used on this map. Next, move to 60 degrees north latitude. What is your expected value, based on the discussion earlier in this chapter, of the length between each degree of longitude? Measure the one degree distance and compare your results to your expected value. One measured value by Kerski was 56 km. The expected value was 56 according to the earlier discussion, it should be exactly half of the equatorial circumference, at 60 degrees north latitude. Does it make sense that if the distance between longitude lines decreases as you move, then the land masses in this projection have to be "stretched" to fit between those lines? It should make sense based on our discussions in this book—the closer to the poles you move, the more the areas become distorted on many maps.

Move back to the Equator and repeat the above process for latitude. First, measure the distance between each successive whole degree of latitude near the Equator, such as between one degree north and two degrees north. How does it compare to the distance between each degree of longitude at the Equator? Why? Second, measure the distance between each successive whole degree of latitude at 60 degrees north latitude. How does it compare to the distance between each degree of latitude at the Equator? Third, measure the distance between each successive whole degree of latitude at 80 degrees north, and compare it to your results along the Equator and at 60 degrees north. The distance between lines of latitude does not vary across the Earth's surface, and so all of your measurements should be around 111 km.

2.11 Summary and looking ahead

All of our geometric analyses in this chapter have been based on Euclidean geometry, assuming Euclid's Parallel Postulate: Given a line and a point not on the line—through that point there passes exactly one line that does not intersect the given line. Non-Euclidean geometries violate this Postulate. What does the geometry of the Earth-sphere become in the non-Euclidean world? We think a bit more about that, and other issues that stretch the imagination, later in this book!

2.12 Related theory and practice: Access through QR codes

Theory

Persistent archive:

University of Michigan Library Deep Blue: http://deepblue.lib.umich.edu/handle/2027.42/58219

From Institute of Mathematical Geography site: http://www.imagenet.org/

Arlinghaus, S. L. and J. Kerski. 2010. MatheMaPics: Educational Research Collaboration. *Solstice: An Electronic Journal of Geography and Mathematics.* Volume XXI, No. 1. Ann Arbor: Institute of Mathematical Geography. http://www.mylovedone.com/image/solstice/sum10/MatheMaPics.html

Arlinghaus, S. L. 2008. Project Archimedes. *Solstice: An Electronic Journal of Geography and Mathematics.* Volume XIX, Number 2. Ann Arbor: Institute of Mathematical Geography. http://www-personal.umich.edu/~copyrght/image/solstice/win08/test/Pirelli07/Pirelli2007.html

Arlinghaus, S. L. and W. C. Arlinghaus. 2005. *Spatial Synthesis: Centrality and Hierarchy.* Volume I, Book 1. Ann Arbor: Institute of Mathematical Geography. http://www-personal.umich.edu/~copyrght/image/books/Spatial%20Synthesis2/

Arlinghaus, S. L. 1990. Parallels Between Parallels. *Solstice: An Electronic Journal of Geography and Mathematics.* Volume I, No. 2. Ann Arbor: Institute of Mathematical Geography. http://www-personal.umich.edu/~copyrght/image/solstice/sols290.html

Practice

From Esri site: http://edcommunity.esri.com/arclessons/arclessons.cfm

Kerski, J. Exploring Measurement. Complexity level: 1, 2. http://edcommunity.esri.com/arclessons/lesson.cfm?id=489

Kerski, J. Exploring Measurement—ArcGIS version. Complexity level: 3. http://edcommunity.esri.com/arclessons/lesson.cfm?id=521

Kerski, J. Distance=Rate × Time with GIS. Complexity level: 3. http://edcommunity.esri.com/arclessons/lesson.cfm?id=519

Kerski, J. Distance=Rate × Time: ArcGIS Version. Complexity level: 3. http://edcommunity.esri.com/arclessons/lesson.cfm?id=527

Transformations
Analysis and Raster/Vector Formats

Keywords: Data (95), map (51), resolution (39), points (32), spatial (30)

They cannot look out far,
They cannot look in deep;
But when was that ever a bar,
To any watch they keep.

Robert Frost
Neither Out Far, Nor In Deep

3.1 Transformations

Mathematics, like musical composition and other fine arts, is purely a human creation. Without us, does it exist? This sort of "meta" question has long interested scholars with multidisciplinary interests. Indeed, does the societal culture and historical epoch in which the predominant mathematics is developed and embedded influence the kind of mathematics that is developed? Again, this question might be studied in many ways: We consider one case here—that of the mathematical relation and selected real-world interpretations. These are displayed in a number of visual formats not merely as curiosities but more significantly for the suggestion they might offer as to why or why not certain types of formal structures get created. It is important to attempt to understand deeper processes such as these: The mathematics we use in the real-world often influences the decisions we make (and vice versa). All data that we use are created, modeled, and maintained based on mathematics. This includes how data are classified, how they are symbolized, what map projection the data are cast in, and even, as we will consider in this chapter, the transformations that are used on the data and the model that is used to represent them on maps and in spatial databases.

Municipal authorities might use demographic forecasts based on curve fitting to guide the direction of urban land use planning. A City Administrator might use a rank-ordered set of priorities to decide how valuable taxpayer funds will be allocated over a period of years to develop (or not develop) infrastructure. The way in which the mathematics is used will influence the outcome of the analysis and, therefore quite likely, the policy that is set in place (Arlinghaus, 1995).

3.1.1 One-to-one, many-to-one, and one-to-many transformations

Much of modern mathematics considers functions as a primary form of mathematical transformation. A *function*, mapping a set X to a set Y permits an element x in X to be sent to an element y in Y, or it permits a number of elements, x_1, x_2, x_3 in X to be sent to a single element y in Y (**Figure 3.1**, left side). In the former situation, the transformation is one-to-one and in the latter it is many-to-one. Functions may be one-to-one or they may be many-to-one transformations. They are "single-valued." Graphically, the idea is represented in **Figure 3.1** (left side). There are also one-to-many transformations. Functions may not be one-to-many transformations. Thus, one-to-many transformations between two mathematical sets are an often neglected class of relationships. When one element of X is permitted to map to many elements of Y, as in x mapping to y_1, y_2, y_3 (**Figure 3.1**, right half) the associated mathematical transformation, that is not a function, is often referred to as a *relation*.

In the Cartesian coordinate system, the same idea may be visualized (**Figure 3.2a–c**). In the case of a function, a vertical line cuts the graph of the function no more than once (**Figure 3.2a** shows a one-to-one function and

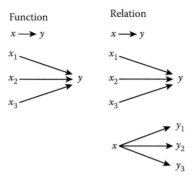

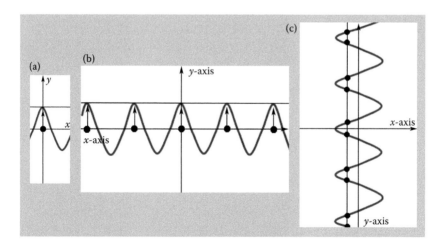

Figure 3.1 *A function requires that each element of X be associated with only one element of Y (although many different elements of X may be associated with the same element of Y). A relation removes this restriction, allowing one element of X to be associated with many elements of Y. Thus, every function is a relation, but not every relation is a function. Based on Arlinghaus, S. L. and W. C. Arlinghaus. 2001. The Neglected Relation. Solstice: An Electronic Journal of Geography and Mathematics. Vol. XII, No. 1. Ann Arbor: Institute of Mathematical Geography. http://www-personal.umich.edu/%7Ecopyrght/image/solstice/sum01/compplets. html.*

Figure 3.2 *(a) One-to-one function. (b) Many-to-one function. Many x-values (black dots) correspond to one y-value (height of horizontal line). (c) One-to-many relation. One x-value corresponds to many y-values (black dots). Based on Arlinghaus, S. L. and W. C. Arlinghaus. 2001. The Neglected Relation. Solstice: An Electronic Journal of Geography and Mathematics. Vol. XII, No. 1. Ann Arbor: Institute of Mathematical Geography. http://www-personal.umich.edu/%7Ecopyrght/image/solstice/sum01/compplets.html*

Figure 3.2b shows a many-to-one function). In a graph that is not a function (not "single-valued"), the vertical line may cut this curve (that is not the graph of a function) in more than one place (**Figure 3.2c** shows such a graph).

The visual display of the difference between function and relation, the many-to-one and the one-to-many, is clear in the Cartesian coordinate system because the ordering of the function from X to Y is clear in our minds. In a coordinate-free environment, such as the world of the applet (a small application that runs tasks within a larger program), all that is evident is the structural equivalence of many-to-one and one-to-many transformations. In **Figure 3.3**, note the stability of the one-to-one transformation as the graphic moves; the many-to-one and the one-to-many never quite settle down to a totally stable configuration. This lack is a function of pattern involving length of edges joining nodes and dimension of the square universe of discourse in which the applets live. In the case of **Figure 3.3**, it may simply be a function of a particular commensurability pattern of edges and underlying raster; nonetheless, the general consideration as to what sorts of configurations exhibit geometric stability is an important one, particularly as in regard for looking for points of intervention into process.

The relation is often ignored in mathematical analyses of various sorts. Perhaps that is because the definite nature of single-valued mappings is regarded as important. Is the world, however, single-valued? We consider a few real-world

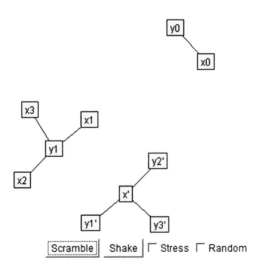

Figure 3.3 *Applets show one-to-one, many-to-one, and one-to-many transformations. Note the structural equivalence between the many-to-one and the one-to-many applets. The printed image is a screen capture of the dynamic applet. Source: Arlinghaus, S. L. and W. C. Arlinghaus. 2001. The Neglected Relation. Solstice: An Electronic Journal of Geography and Mathematics. Vol. XII, No. 1. Ann Arbor: Institute of Mathematical Geography. http://www-personal.umich. edu/%7Ecopyrght/image/solstice/sum01/compplets.html*

situations in which relations can be observed to be the underlying conceptual force.

3.1.1.1 Postal transformation

A simple, convenient example often given to students studying functions for the first time is the following postal example. Suppose one is given a set of hard-copy handwritten letters in envelopes that are to be sent through the conventional US Postal Service network by regular first-class mail.

- My letter can be sent to a single address (one-to-one).
- My set of three different letters can be sent to a single address (many-to-one).
- My one letter cannot go to three different addresses (not one-to-many).

Some might argue that the invention of the printing press permitted one page to go to many. Yet, there is variation from page to page—there are ink splatters, broken type, and so forth.

Still others might assert that photocopying of a page will enable one letter to go to many different addresses, as long as the original as distinct from the rest is not included—hence the rise of junk mail. Someone else might argue, however, that any two photocopies differ from each other on account of diminishing the amount of toner available for copies later in the process.

Further, if one considers virtual messages, rather than hard copy messages, then a single e-letter can be sent to a single address (one-to-one), a set of three different e-notes can be sent to a single address (many-to-one), and a single note can go simultaneously to three different addresses (one-to-many). The electronic revolution of our "information age" offers a true postal transformation from the functional to the relational.

Perhaps a common theme in all these refinements of argument will be that to move from one style of mathematical transformation to another in the real-world requires some sort of underlying real-world transformation through invention, revolution, or other remarkable event. Hence, we argue that the printing press, the photocopying machine, the computer, and e-mail were not mere inventions but were transformational technologies. They transformed workflows and processes in entire organizations and in the greater society. Electronic journals, delivered over the Internet, take advantage of this one-to-many relational capability.

3.1.1.2 Home ownership

As we look around our environment today, such as a typical community in the United States of America, we see a variety of dwelling types and a variety of dwelling ownership.

- One-to-one ownership: One family owns a single parcel of land, often in a suburban area and elsewhere when land is plentiful and land values are relatively low.
- Many-to-one ownership: Many families own a single parcel or building. This style of ownership is often "condominium" or "cooperative" ownership. In the landscape, it is evident mostly in more densely populated areas or where land values are relatively high.
- One-to-many ownership: One family owns many residences. This particular situation, not represented as a mathematical function but only as a relation, is perhaps not as common as the two above. Typically, one might expect families with excess wealth to own more than one residence. Our colleague John Nystuen asked where, in the situation with multiple residences, such individuals cast votes. We explore the dynamics of that situation below.

3.1.1.3 Composition of transformations

If one were to map the relations listed above for home ownership, a figure similar to **Figure 3.3** would be the result. When voting is added on, the situation becomes more complicated, given that voting is done and counted locally and not nationally.

- In the one-to-one situation, the homeowner registers to vote from his or her single address and there is no difficulty counting the vote.
- In the many-to-one situation, the homeowners register to vote from their unique, single address and there is no difficulty counting the vote. All go to the same polling place to vote.
- In the one-to-many situation, however, a person who owns property in Michigan and in Florida, for example, might attempt to vote in two places even though he/she is only entitled to one vote.

Figure 3.4 shows that when only a single vote is cast, as it should be, the system remains closed, bounded, and manageable (in some sense). In **Figure 3.4a**, voter x owns three residences, y_1, y_2, and y_3 and casts the one legal vote, z, to which he/she is entitled. Voter a owns three residences, b_1, b_2, and b_3 and casts one legal vote c_1 (from residence b_1) and two illegal votes (from residences b_2 and b_3), c_2, and c_3. Note that the legal case is visually manageable in some sense while the illegal case sprawls across the map and is more difficult to track. When more than one vote is cast, the system may rapidly fall out of order, especially when there are thousands or hundreds of thousands of people who own more than one residence from which they might vote. When home ownership and voting become more complicated (**Figure 3.4b**), the closure and sprawl issues, referred to above, become more evident. Some sort of nationalized database on voter registration and residency makes the problems easier to deal with.

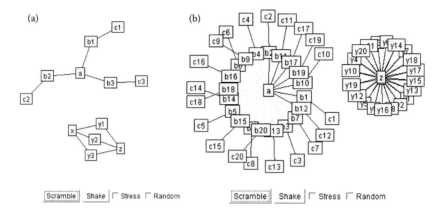

Figure 3.4 *(a) Simple case of home ownership and voter registration. (b) Complex case of home ownership and voter registration. Source: Arlinghaus, S. L. and W. C. Arlinghaus. 2001. The Neglected Relation. Solstice: An Electronic Journal of Geography and Mathematics. Vol. XII, No. 1. Ann Arbor: Institute of Mathematical Geography. http://www-personal.umich.edu/%7Ecopyrght/image/solstice/sum01/compplets.html*

What is important in this case is the composition of transformations: One transformation followed by another. In this case, the two transformations involve home ownership followed by voting. When the first is a function, the composition works well in the real-world interpretation. When it is only a relation, there is room for serious manipulation that was not present in the functional characterization. Extra care is appropriate when composing transformations.

3.1.1.4 Other one-to-many situations

- Quadratic Probing is used to resolve collisions, which are situations where many pieces of data are initially assigned the same data location. This is an undesirable one-to-many situation, as one data location cannot store many pieces of data.
- One copy of material may be photocopied according to certain legal constraints; however, when a "single" copy is sold to many, there may be copyright infringement; one-to-many becomes problematic when the transformations of "single copy" and "selling" are composed.
- In the state of Michigan, the same driver license number can be assigned to two or more individuals. Two individuals with the same month and day of birth, the same first and middle names, the same first letter of last name, and initial parts of the last name the same, have the same Michigan driver license number. For example, James Edward Smithsonian, born July 1, 1900, and James Edward Smithson,

born July 1, 1950, would have the same Michigan driver license number. The first position would hold the letter "S" from the last name; the next three digits would be based on the first three consonants (mth) in the last name; the next three digits would be based on the first name (James), so about a third of the way through the three digit numbers from 000 to 999—maybe somewhere in the high 300s; the next three digits would be based on the middle name (Edward), maybe in the high 100s; and the final three would be based on the birthdate of July 1, so maybe about 500.

- In the past, there were stories of a local resident with two different driver licenses under two different names, an impermissible one-to-many relation.
- Marital relationships (men and women):
  - One-to-one: One woman is married to one man (monogamy: Customary practice in the part of world predominant in mathematical development).
  - Many-to-one: Many women are married to one man (polygamy: Known, but not customary; some examples exist in European history, in the idea of a harem, and in certain religions).
  - One-to-many: One woman is married to many men. Various cultural taboos might lead one to ask what kind of mathematics would have been developed in the 20th century had it been done so predominantly by a society in which fundamental societal relationships are one to many. The broader societal environment in which mathematics develops may well influence what mathematics gets created. As our human environment changes, do you see changes that might produce substantial shifts from one transformation to another?

Finally, we note that the ideas of one-to-many and many-to-one might be combined, as many-to-many. From the standpoint of application and education what is important here is to see that:

- In the world of GIS, all of one-to-one, many-to-one, one-to-many, and many-to-many concepts arise in mapping;
- Thus, the world of mathematical education needs to respond by offering extensive training not only about "functions" but also about the often neglected "relation."

3.1.2 Geoprocessing and transformations

Contemporary GIS software contains small programs designed specifically to deal with one-to-many situations. Attributes for mapped features, such as whether streams are perennial or intermittent, or the depth and magnitude of earthquake epicenters, for example, are stored in a series of tables in a geographic database, or "geodatabase." These tables can be related in several different ways. Establishing relationships between attribute tables is useful so that the relationships can be mapped and analyzed. For example, in Esri's ArcGIS for Desktop, most map layers contain an attribute table. The map can be thought of as the "G" part of GIS, while the attribute table can be thought of as the "I" part of GIS. The "S" part is the relationship that links each feature (such as a river, a county, a hurricane track) with its associated attributes (such as whether the river is perennial or intermittent, the name of the county and its 2010 population, and the intensity of the hurricane, respectively). This relationship is usually maintained via a common feature ID linking the feature to its attributes. If this relationship is broken, not only will the features not have attributes, but the database is broken and the features will not even be able to be mapped.

However, additional relationships can also be established within the geodatabase. A one-to-one or many-to-one relationship can be established between a map layer's attribute table and another, separate, external table. An example of this happening via a one-to-one relationship is for weather stations. Each station is represented by a point on a map. Each point is represented by a record, or row, in its associated Table A. Table B contains the weather data, such as temperature, humidity, wind speed, and other variables for each station. Table B contains multiple columns, or fields, for each station, but each station is still represented by one row in the table. A common ID establishes a one-to-one relationship between the two tables, allowing the analyst to map the weather stations and analyze the spatial pattern of the weather variables on that map.

At other times, a many-to-one relationship is needed. For example, suppose that a particular polygon map layer shows types of land use in an area. The land use layer's attribute table only stores a land use code. A separate table stores the full description of each land use type. Joining these two tables together establishes a many-to-one relationship because many records in the layer's attribute table are related to the same record in the table of land use descriptions. The analyst can use the more descriptive text when generating the legend for the map. This descriptive table is often referred to as a "lookup table."

In a GIS, tables can be joined together to perform the activities above, but tables can also be related without performing the join. Relating tables in ArcGIS for Desktop is done with a "relate" function, which simply defines a relationship between two tables. The associated data are not appended to the layer's attribute table as they are with the process of "joining." Instead, the related data can be simply accessed when the analyst works with the layer's attributes.

For example, if you as the analyst select a building, you can find all of the tenants that occupy that building. Similarly, by selecting a tenant, you can find what building the tenant resides in. The tenant may reside in several buildings, as in the case of a chain of stores in multiple cities and in multiple shopping centers. This is an example of a many-to-many relationship. In summary, you should join two tables when the data in the tables have a one-to-one or a many-to-one relationship, and you should relate two tables when the data in the tables have a one-to-many or many-to-many relationship.

3.1.3 QR codes

Bar codes, and their two-dimensional counterpart, QR codes, permit a transformation from print to electronic format. We use them extensively throughout this book to explain concepts with animations, applets, extra color, and so on either when those concepts cannot be included using standard print media or when they are difficult or expensive to include in a different format. The Smartphone, loaded with an appropriate QR code reader, performs the transformation. A few years ago, we might not even have dreamt of this possibility; now this transformation is commonplace: QR codes appear in advertising on buses or subways, on hospital wrist bands, on cemetery markers (W. E. Arlinghaus, 2011), and a host of unexpected spots. Where have you seen them? Where might you see them? What sorts of applications might you consider for them?

What is important is that one QR code might link to one location or many different QR codes might link to a single location. The transformation is a function. A single QR code, however, may not link to multiple locations—a good thing! For example, a memorial QR code on a cemetery marker might link to obituary text; another QR code posted in a local newspaper might also link to the same obituary. That pattern would be fine; however, it would not be fine to have the single QR code on the cemetery marker link to obituaries of two different people! What sorts of far-reaching transformations might you imagine?

3.2 Partition: Point-line-area transformations

3.2.1 Buffers

Often in spatial analysis, we seek to determine proximity—how far things are from each other. Zones of proximity often involve a GIS concept known as "buffering." Around a point, a buffer of a specified distance becomes a circle. Around a line segment, a buffer becomes a sausage-like shape, while around a polygon, a buffer takes the shape of the polygon but appears rather "inflated" or "puffy."

Whether one considers accessibility to railroad service using linear buffers of tracks, counts population in buffered bus routes, or selects minority groups from within circular buffers intersecting census tracts, the buffer has long served, and continues to serve, as a basis for making decisions from maps. Buffers have a rich history in geographical analysis—long before the advent of GIS software. Mark Jefferson (Jefferson, 1928) rolled a circle along lines on a map representing railroad tracks to create line-buffers representing proximity to train service and suggested consequent implications for population patterns in various regions of the world. Julian Perkal and John Nystuen saw buffers in parallel with delta-epsilon arguments employed in the calculus to speak of infinitesimal quantities (Nystuen, 1966; Perkal, 1966). Jefferson's mapping effort in 1928 was extraordinary; today, buffers of points, lines, or regions are easy to execute in the environment of GIS software. To paraphrase Faulkner (1949): "Good ideas will not merely endure, they will prevail."

3.2.2 Buffers build bisectors

Buffers also offer an interesting link to geometry: Buffers serve as bridges in the realm of spatial mathematics; they can be used to create the classical Euclidean construction of a perpendicular bisector. **Figure 3.5a** illustrates Euclid's classical ruler and compass construction for drawing a perpendicular bisector separating any pair of distinct points:

- Given O and O' in the plane.
- Draw a segment joining O and O'.
- Construct two circles, one centered on O and the other centered on O', each of radius greater than half the distance between O and O'. The radii are to be the same.
- Label the intersection points of the circles as A and B. Draw a line through A and B. This line is the perpendicular bisector of $|OO'|$.

Buffering the points O and O' at a sequence of distances in mapping software (ArcGIS), and then joining the intersection points of the pairs of equidistant

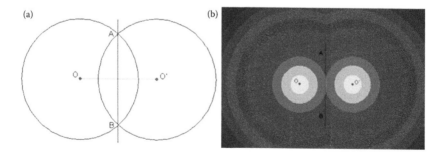

Figure 3.5 (a) Construction of perpendicular bisector, AB, of |OO'| using ruler and compass. (b) Buffering (Euclidean style) the points O and O' at a sequence of distances in mapping software, and then joining the intersection points of the pairs of equidistant buffers, also generates the bisector AB. Source of base map: Esri software. Source: Arlinghaus, S. L. 2001. Base Maps, Buffers, and Bisectors. Solstice: An Electronic Journal of Geography and Mathematics. Vol. XII, No. 2. Ann Arbor: Institute of Mathematical Geography. http://www-personal.umich. edu/%7Ecopyrght/image/solstice/win01/sarhaus.

buffers, also generates the bisector AB (**Figure 3.5b**). The GIS notion of buffer, or proximity, captured the Euclidean construction of perpendicular bisector.

3.2.3 Buffers build bisectors and proximity zones

If there are more than two points in a given distribution, and if one wishes to find perpendicular bisectors within the whole set, the matter can quickly become tedious. Mapping software again offers a quick and accurate way to calculate bisector positions. We illustrate the conceptual basis on which this might be done using circular buffers around a distribution of 25 Canadian cities (viewed as points). **Figure 3.6** shows a pattern of circular Euclidean buffers surrounding these points. The underlying multi-colored mesh of polygons arises from the construction of perpendicular bisectors associated with these buffers. The bisectors are edges of the polygons, and the polygons serve as proximity zones. The reader may find other names used for such zones or

Figure 3.6. *Buffers build bisectors and proximity zones around a set of points and these bisectors in turn create proximity zones. QR code links to an associated animation. Source: Arlinghaus, S. L. 2001. Base Maps, Buffers, and Bisectors. Solstice: An Electronic Journal of Geography and Mathematics. Vol. XII, No. 2. Ann Arbor: Institute of Mathematical Geography. Source of base map: Esri software. http://www-personal.umich.edu/%7Ecopyrght/image/solstice/win01/sarhaus*

related ideas: Thiessen polygons, Dirichlet regions, Delaunay triangulations, and others (Coxeter, 1961a; Kopec, 1963; Rhynsburger, 1973; Thiessen and Alter, 1911). Each zone contains exactly one Canadian city and all territory within that zone is closer to that city than it is to any other city. There are a number of subtleties (such as dissolving arcs within buffers or splitting polygons) in building the map in **Figure 3.6** and some of these are shown in the animation associated with the QR code in **Figure 3.6**. Euclidean buffers such as these are generally reserved for studies more local than this one. The point here is to see the connection between the two-dimensional Euclidean construction for perpendicular bisectors and the GIS creation of buffers as concentric circles. The reader will have a chance to practice with buffers of various kinds (and related ideas) later in this chapter.

3.2.4 Base maps: Know your data!

One of the themes running through this book is the importance of under-standing your data. It is particularly relevant to our discussion here. In this example, note that even though the proximity zones extend beyond the national boundary of Canada, points for which there are data exist only within the boundaries of Canada (dotted white in **Figure 3.6**). Thus, within most proximity zones, some points have associated data while others do not. Hence any average (or other statistical) values across an entire zone should be viewed with caution. Often, extrapolated data sets that lie far from established data points cause problems. And the zones beyond the data points should be viewed with the most caution of all. Know your base maps! Know your data!

3.3 Set theory

While this material is clearly based on a single ruler and compass construc-tion of Euclid (that of perpendicular bisector), it is also based on elements of set theory—a foundational branch of mathematics that systematically stud-ies collections of objects (Hausdorff, 1914). The pair of intersecting circles in **Figure 3.5a** partitions the space that contains them into sets of points: All the points in one of the circles; all the points in the other circle; and the points outside the two circles. The space containing this configuration is the universe of discourse. Because the circles intersect, there are other ways to look at this partition. Consider the circle on the left as A and the one on the right as B. The points within both A and B are said to be within its intersec-tion, denoted $A \cap B$. The points in either A or B or both is called the union of the two sets, denoted $A \cup B$. The points outside are in the complement of $A \cup B$. The configuration in **Figure 3.5a**, described in set-theoretic terms, is a Venn diagram on two circles (Venn, 1880). Try to tie some of these con-cepts more closely with the examples given above. How does the idea of complement relate to the idea of having concern about accuracy away from established data points? To look more deeply at this topic, follow some of the links at the end of this chapter or read references cited at the end of the book.

Set theory terminology also has linkages with standard English language terminology. It is important to note what these are because mapping software uses the mathematically assigned, set-theoretic, meanings for common words. The word "intersection" corresponds to the word "and." That association appears straightforward to most. The word "complement" corresponds to the word "not." That too appears straightforward. Where there is difficulty for some is in the translation of the word "union" (sometimes called "join"). The word "union" or "join" corresponds to "or," where "or" means "either one or both." This form of "or" is said to be "inclusive." In some forms of common usage, "or" may mean "either one or the other but *not* both." That form of "or" is said to be "exclusive." In set-theoretic usage, and therefore in mapping usage, the inclusive "or" is the one that is used. This fact is important because some of the wizards that offer choices for spatial analysis assume that the user knows about the use of "or" (or "union" or "join") in an inclusive context. If one does not know, then it can be quite difficult to figure out why the results, of what appeared to be otherwise correct procedure, have become skewed. Remember—mapping software is built on mathematics!

3.4 Raster and vector mapping: Know your file formats

A Geographic Information System (GIS) is designed to present all types of geographical data. It rests on base maps and an underlying table of information about the points, lines, and areas within the base maps. This table is referred to as an "attribute" table. GIS technology elevates the art and science of mapping from a static one to a dynamic one. In GIS, a change in the database produces a corresponding change on the map. Similarly, a change on the map produces a corresponding change on the database.

Gone are the days when a change in a place name necessitated redrawing the map! When maps are made on paper, they are made in a static mode. If an error is made while one is using India ink to draw a country outline, a serious problem arises. In times past, we pulled out a razor blade to carefully scrape the ink off the surface of our expensive high-rag-content paper. Because part of the finish of the paper also got removed, thereby causing ink to smear when reapplied to the scraped spot, we restored the finish using a special restorative powder. Blowing the powder off could result in the introduction of water, causing yet another set of problems. So, we used nice camel hair brushes that did not shed to remove the excess powder. Finally, we could consider inking the spot again, perhaps with extra care so as not to create yet another error on the newly "fixed" area. During the era where maps were etched with scribing needles onto pieces of film, that were later exposed to create maps, paint was applied to the erroneous line, allowed to dry, and the line was re-etched through the paint. Often, the best solution was simply to throw the film away and start over again.

GIS not only permits dynamic response to such changes, but it also offers hundreds of algorithms that allow the user to have a wide-ranging control of maps and data with the freedom to customize them creatively at will. Thus, understanding of what one is doing is critical!

The introduction of digital mapping caused a revolution in mapping: The process became dynamic. A change in the database produced a corresponding change on the map, and vice versa. Errors on a map could be corrected almost as easily as errors in a text document. There was no need to start over. Production improved in quality and speed. What remains critical in both approaches, however, is to understand the concepts involved and to think about what we are doing as we create maps.

So, drawing from the classical analogy, what is the "paper" for a digital map? Clearly it is the computer screen. Having the map there is what makes it possible for mapping to become dynamic. To be effective, however, we need to understand the mechanics of the computer screen as well as we understood paper types and paper finishes.

The computer screen is composed of a set of small squares, called pixels, each of which holds a small part of the image—like pieces in a jigsaw puzzle of tiny squares. Generally speaking, the smaller the squares, the sharper the image; sharper images are images of high resolution. Thus, we can imagine assigning location on the computer screen in terms of pixel location (the third pixel down from the top and the seventeenth pixel over from the left). That sort of assignment is like using parallels and meridians. It is relative to the pixel size. Instead, one might also use mathematical coordinates to assign location independent of pixels. This approach using standard mathematics is much like introducing standard circular measure to replicate results on the sphere (see **Chapter 1**).

When maps or other images are created, they might be composed of a set of instructions on how to form the shapes with specific numerical values; they are made up of mathematically-defined paths as line segments and curves (**Figure 3.7a**). This sort of image is called a vector image; it does not use pixels and is based on standard mathematics. Or, images can be composed of a set of pixels with each pixel having a specific value or color. This sort of image is called a raster image (**Figures 3.7b** and **c**). There are merits and drawbacks to maps based on images formed in either approach. Bear in mind, though, that the computer screen is itself composed of pixels and therefore all images, no matter how constructed, will appear as raster-like on the computer screen. How they print out is a different matter.

For vector images, in the associated vector database, each image feature has an attribute in the underlying table. Categories of data can be suppressed by the software producing the map, so that there is capability to show denser data by zooming in. The image can be rotated and have text remain upright.

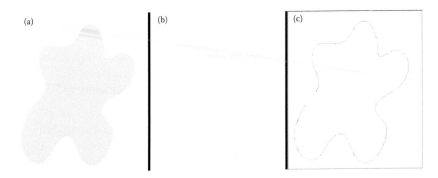

(a) *(b)* *(c)*

Figure 3.7 *(a) Vector image, (b) grid (raster) overlain on curve, and (c) raster image in yellow.*

Raster images look and feel like a paper map; this format can be updated frequently using raster patches, but this format does not have the capability to show denser data when zooming in. When the image is rotated the text remains fixed and is therefore not upright.

These observations led us to attempt to encapsulate the general merits and limitations of raster and vector images.

- Raster image: Merits.
    - Photographs and .jpg files are raster images. If a raster image is to be aligned with a map, it makes sense to choose a raster representation for the map so that the pixels will overlay correctly throughout the image.
    - These images are easy to edit (in a variety of ways, including the editing of the color's hue and saturation). They can display fine nuances in light and shading. They are often a preferred file format for conventional publication not concerned with scalability.
    - These images have the option of applying filters for visual enhancement such as drop shadows, inner or outer glow, and so forth.
- Raster image: Limitations
    - These images have a set resolution (coming from pixel base) and will become fuzzy (pixelated) when enlarged. These images are not truly scalable.
    - When zoomed in on, curves on these images may become jagged in appearance, due to the corners on the pixels.
- Vector image: Merits
    - Can be scaled to any size and will retain its sharpness.
    - Can zoom in on the screen and see smooth, rather than jagged, edges independent of scale.
    - Useful for highly geometric images, such as logos, for example, that can be perfectly scaled to fit on media of highly variable size (stationery, business cards, etc.)

- Vector image: Limitations
  - Often these images are difficult to align with raster photographs, such as aerial images.
  - Vector formats may be more difficult to edit than raster formats. Extra filters may not be available.
  - Vector images must be converted to raster images in order to display on a website.

For the purposes of mapping, there are two very important issues:

- If a photograph is to be combined with a map, bear in mind that alignment of the photograph with the map will need to account for the raster on which the photograph is based.
- If scalability is important, bear in mind that vector images are the only truly scalable images.

No one format is "best." Taking advantage of multiple options requires knowledge and planning. Think about what is important as a final product before embarking on mapping: If scalability is critical, as it often is, choose a vector format. If combining a photo in with a map is important, consider a raster format. The decision is yours. Again, there is no "perfect map" ... not in paper format, and not in digital format. The "Mapmaker's" quest for perfection cannot be realized!

3.4.1 Representing the Earth using raster and vector data

Both vector and raster formats have long been used to represent data in a GIS. Vector data are used to represent mostly discrete phenomena such as points (such as water wells or volcanoes), lines (such as rivers or pipelines), and polygons (such as lakes or census tracts). Once again, scale matters: At a small scale, cities or rivers might be best represented as points, but at a larger scale, as polygons. Raster data are used to represent continuously changing phenomena, such as elevation or snowfall depth. However, rasters are also used to represent thematic map layers, such as types of soil, zoning in an urban area, or land use. And even more commonly, rasters are used to represent remotely sensed imagery, including aerial photographs and satellite images. Rasters are used with imagery not only to represent the reflectivity in the visible band of the electromagnetic spectrum, but also in bands that the human eye cannot see, such as radar and infrared. These images are used to detect ore bodies, to map urban infrastructure in 3D, to analyze the response of forests to pine beetle infestation, and much more.

Resolution in both the vector and raster data are tied to the scale at which that data set was collected. Simply because nowadays one can zoom in to a fine resolution (large scale) on the screen where one is running GIS software does not mean that the data in that GIS has been collected at that scale. For example, one can zoom in to a scale of 1:1000 in a GIS, but the data may have been collected at 1:24,000 scale. Owing to modern, fine screen resolution, the features may look like they are in their "true" locations. However, a buffer zone of accuracy, if it were available, would show rings around each feature, such as buildings and water wells. From those rings, the data user would be astounded to see the many locations possible for these features. Therefore, knowing the spatial resolution of the data is key to making wise decisions with it. One should not make decisions at a finer scale than that at which the data were collected.

3.4.2 Vector data resolution: Considerations

Vector data resolution has to do with the positional accuracy of the features represented, and the attributes of those features, the "G" and the "I" components of a GIS, respectively. Many digital data sets in use today were originally produced in paper map form and scanned or otherwise converted into digital formats, and thus carry with them the accuracy standards of their original maps. They are no more accurate in the digital world than they were in the analog world. Standards for paper and digital mapping indicate the tolerance for error; they do not imply that maps or the digital data produced from maps are perfect.

For example, as we have seen earlier and will see again later, one of the first things done with mapped data is to choose what projection they should be cast in. Projections by their very nature are distortions of reality, in area, distance, direction, and/or shape. In terms of horizontal accuracy, consider the US National Map Accuracy Standards: For maps on publication scales larger than 1:20,000, not more than 10 percent of the points tested shall be in error by more than 1/30 inch, measured on the publication scale; for maps on publication scales of 1:20,000 or smaller, 1/50 inch. These limits of accuracy shall apply to positions of well-defined points only. In terms of vertical accuracy, the National Map Accuracy Standards state "Vertical accuracy, as applied to contour maps on all publication scales, shall be such that not more than 10 percent of the elevations tested shall be in error by more than one-half the contour interval." (Read more about the National Map Accuracy Standards on http://egsc.usgs.gov/isb/pubs/factsheets/fs17199.html.) Again, this idea implies that some features will in fact be off by more than these amounts. Thus, when we think about map accuracy, error must be managed; it cannot be eliminated. In terms of the attributes in a vector data set, resolution has to do with such things as the number of attributes a data set contains, and how many different categories of each attribute it contains (such as the Anderson land use classification scheme that contains several resolutions of completeness (Anderson, 1976)). The data user must determine whether the accuracy of location and attributes can serve the project requirements; in other words, whether any given data set is "fit for use." It is the data producer's responsibility to document the accuracy and completeness for any data set in the set of metadata provided to the user.

3.4.3 Raster data resolution: Considerations

Four types of raster data resolution compose the "quality" or "scale" of raster data. The first of the four types is spatial resolution. Spatial resolution refers to the amount of land area that is represented by each pixel. A data set with a 0.5-meter spatial resolution is finer than a data set with 10-meter spatial resolution. Higher spatial resolution means that there are more pixels per unit area, and is a result of how much of the Earth's surface the sensor on the airplane or satellite "sees" with each image captured. The part of the Earth's surface being sensed is known as the Instantaneous Field Of View, or IFOV. The size of the area viewed is determined by multiplying the IFOV by the distance from the ground to the sensor. If a feature on the ground is greater in size than the resolution cell, then that feature can be detected. Conversely, if the feature is smaller than the resolution cell size, it cannot be detected. Therefore, a 10-square-meter lake would not be detectable from a sensor with a 30-square-meter spatial resolution.

Spectral resolution refers to the number of spectral bands a data set has and thus is only applicable to rasters that are created from sensors capable of recording more than one band. This is usually confined to some aerial imagery and most satellite imagery. It depends wholly on the ability of a sensor on a satellite or aircraft to distinguish between wavelength intervals in the electromagnetic spectrum. A higher spectral resolution corresponds to a narrower wavelength range for a particular band. Where a grayscale orthophotograph has a low spectral resolution because it captures data from across the entire visible range of the spectrum, hyperspectral sensors collect data from up to hundreds of very narrow spectral bands.

Temporal resolution or the "revisit period" has to do with the frequency with which images are captured over the same place on the Earth's surface. Therefore, a sensor that captures data once every three days is said to have a higher temporal resolution than multispectral scanners on the first Landsat satellites, which could revisit the same place only every 18 days. Temporal resolution depends on the satellite's latitude, its physical capabilities, and the amount of overlap in the image swath.

Finally, radiometric resolution represents the ability of a sensor to discriminate between very slight differences in reflected or emitted energy. Image data are generally displayed in a range of gray tones, with black representing a digital number of 0 and white representing the maximum value. Image data are represented by positive numbers from 0 to one less than a selected power of two. This corresponds to the number of bits used for coding numbers in a computer. All numbers in a computer are stored as binary numbers—0s and 1s. The number of levels available depends on how many computer bits are used in representing recorded energy. If a sensor used 8 bits to record the data, there would be 2 to the 8th power, or 256 available values, from 0 to 255. However, if only 4 bits were used, then only 2 to the 4th power, or 16 values, would be available, and the resulting image would have less radiometric resolution. A Landsat band is typically 8-bit data, while a GeoEye IKONOS band is typically 11-bit data and therefore has a higher radiometric resolution.

To select an appropriate spatial resolution, just as when considering other spatial data for use, there is no perfect solution. One guideline is to choose a resolution that is a factor of ten times finer than the size of the features you need to identify. For example, if you want to visually delineate features with a minimum size (referred to as the "minimum mapping unit") of one square kilometer (1 kilometer × 1 kilometer on each side, which is 1,000,000 square meters, or 1000-meter resolution), a 1000/10, or 100-meter spatial resolution is probably sufficient. To identify tree crowns that are three meters by three meters in size, you would need to select a one meter or finer resolution. A mathematical relationship exists between the map scale and the image resolution. The rule is to divide the denominator of the map scale by 1000 to get the detectable size in meters. The resolution is one half of this amount. "Of course the cartographer fudges. He makes things which are too small to detect much larger on the map because of their importance. But this cannot be done for everything so that most features less than resolution size get left off the map. This is why the spatial resolution is so critical" (Tobler, 1987, 1988).

To determine an appropriate mapping scale from a known spatial resolution, use the following formula: Map Scale = Raster resolution (in meters) × 2 × 1000. The number "2" in this equation is the minimum number of pixels required to detect something. Thus, if you have an image of 1-meter resolution, you can detect features at a map scale of 1:2000 using this formula: $1 \times 2 \times 1000$. The spatial resolution needed to detect features at a map scale of 1:50,000 is approximately 25-meter [50000/ (1000*2)] resolution. This may be helpful if you need to acquire satellite imagery to digitize vector data layers against, or you already have imagery and need to know the scale of map in which it can be used.

3.4.4 Determining if a data set is fit for use

Fundamental to the effective integration of geography and mathematics is how to determine whether a data set is fit for use in addressing a problem or issue. Five measures of accuracy exist that should be considered when determining whether the spatial data are fit for one's own use. The first is positional accuracy: How close are the locations of the objects to their corresponding true locations in the real world? Will this be sufficient for your needs? The second is attribute accuracy: How close are the attributes to their true values? The third, logical consistency, asks such questions as: If you used this data set with other data sets, will its spatial characteristics and attributes cause strange juxtapositions or illogical associations, such as a road that is also a canal? Does every area have a label point? Is the data set consistent with its definitions? The fourth is completeness: Does your data set completely capture all of the features that you need? The fifth component is lineage: Does the history of the data set, including who created the data set, how the data set was created, when the data set was collected, and what methods were used to collect the data set, meet your requirements?

Recall from our earlier discussion in this book that accuracy is not the same as precision. A data set's accuracy is the closeness of results, computations or estimates to true values, or values accepted to be true. A data set's precision, on the other hand, is the number of significant digits (decimal places) contained in a data set's measurements. Are your data precise and accurate enough for your needs? If not, can you live with your data? More important, can your end-users live with the products or decisions that are derived from the data you used?

In the following activities section, you will have the opportunity to practice concepts from this chapter including buffering and overlay operations. We will begin with two buffering activities that you can use on the Internet and follow it up with in-depth overlay operation that will enable you to locate a business in a metropolitan area.

3.5 Practice using selected concepts from this chapter

3.5.1 Drawing buffers from different types of features

As we have discussed, buffers are an essential tool in spatial analysis. Using a web browser, access the following buffering toolkit that uses ArcGIS Online: http://resources.esri.com/help/9.3/arcgisserver/apis/javascript/arcgis/demos/ utilities/util_buffergraphic.html. Click on the line button but use the same start and end node for your line, so that it is, in essence, a point. What shape does your resulting buffer have, and why? Is the buffer 25 miles in diameter or 25 miles in radius? Why? The buffer should be a circle, because a buffered point is a circle. It should have a radius of 25 miles because every point within the circle is 25 miles or less from the original drawn point.

Next, create a polyline (essentially, a line with multiple segments in GIS terminology). What is the shape of the resulting buffer, and why? No matter what shape of polyline you draw, your buffer should be a "thicker" or "puffier" version of that line, resembling a snake or a sausage. Next, create a polygon. What is the shape of the buffer around a polygon, and why? Your buffered polygons will have the shape of the original polygon, with an extra amount of space outside the boundaries of the original polygon, making the polygon look "puffy."

What questions should you ask about the underlying map projection when creating buffers, and why does the map projection matter? When creating buffers or anything else involving distance, angles, shapes, and areas, questions about the map projection should always be first and foremost. The reason is that the map projection affects not only the appearance of the buffers, but how the buffers are computed in the first place. The distances that are used to create them depend on the underlying map projection. This will be evident in the next section as you change the scale and look at larger areas.

Zoom out until you see most of the Earth. Change the buffer distance to 100 miles or 100 kilometers. Draw a polyline near the Equator, and compare it to a polyline that you draw near the North Pole. Next, draw a polyline from northern Canada to the Equator. What do you observe? Why does the buffer distance change with latitude? You should see that the buffer distance appears to be wider as you approach the poles, and narrower as you approach the Equator. The buffer distance is the same across the planet, but because the map is cast onto the Web Mercator projection, in order to remain true to the projection, the buffers have to widen to be shown accurately near the Poles. Along with the buffers, you should notice that the underlying land masses also appear to widen and become misshapen.

3.5.2 Geodesic versus Euclidean buffering

As we have discussed, like all spatial analytical tools, buffers fundamentally depend on the mathematics behind them, and specifically, the shape of the Earth. Geodesic buffers account for the actual shape of the Earth as an oblate spheroid in their calculations. Euclidean buffers measure distances in a two-dimensional Cartesian plane. Euclidean buffers work best when analyzing distances around features that are concentrated in a relatively small area, in a projected coordinate system. Using a web browser, access the Geodesic Buffering web GIS application on: http://resources.arcgis.com/en/help/flex-api/samples/index.html#//01nq0000002q000000.

Click anywhere on the map to generate a line. What shape do the resulting buffers have, and why? The short line segments that you generate create buffers that are oval in shape, because they represent shapes that enclose areas within 1000 kilometers of the line segments. **Figure 3.8** illustrates geodesic and Euclidean buffers around Detroit, Michigan. Move closer to the North Pole, redraw a line, and then repeat closer to the Equator. Why is the geodesic buffer (shown in red) larger than the Euclidean buffer (shown in blue)? The geodesic buffers are larger because they consider the Earth as a curved object, whereas the Euclidean buffers treat the Earth as a flat plane. Where are the two buffers closest in size to each other? Why? The two buffers are most similar near the Equator and are quite different near the poles. Interpretation problems can ensue when performing a Euclidean buffer on features stored in a projected coordinate system where the map projection distorts distances, angles, areas, and the shapes of features. The danger of using the wrong kind of buffer will become obvious when we discuss a notorious example regarding areas within range of a certain country's missiles later in this book.

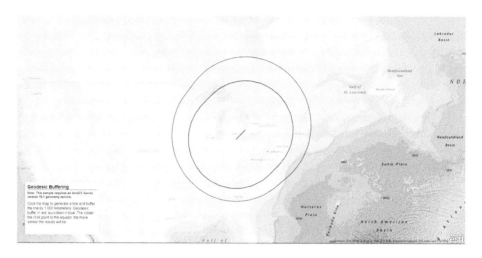

Figure 3.8 *Geodesic (in red) and Euclidean (in blue) buffers around Detroit, Michigan. Source: Esri base map.*

In the next section, you will use GIS to site the optimal location for a high-speed Internet café in Denver. The activity includes buffering features as discussed in this chapter, as well as querying, computing summaries, classification, and other techniques central to spatial mathematics.

3.5.3 Siting an Internet café in Denver

The activity involving the siting of an Internet café in Denver, linked near the end of this chapter, uses spatial analysis in a GIS environment to (1) download, format, and use data from the Esri Data depository, and (2) site an Internet cafe in Denver using the downloaded data, considering demographics, traffic volume, and proximity to educational institutions. This exercise is more advanced than are many of the others in this book. Readers with previous experience using GIS software should benefit from that experience. To complete the entire exercise requires about three hours and we assume that the reader has access to ArcGIS 9 or later. In the body of the text, here, we present a few of the elements of this activity and encourage the reader to participate in the full activity, now, or after reading more of this book. Indeed, we shall refer to it, and present other parts of it, later.

Locating any business in a metropolitan area requires a consideration of spatial factors, and therefore, GIS is an essential tool to bring analytical capabilities to the problem. These include: Downloading, formatting, and understanding Census TIGER spatial data and US Census demographic data, joining attribute tables, tabular and spatial sorting and querying, overlay and proximity analysis, creating map layouts, and most important, solving a problem based on spatial analysis.

Problem: You are a new franchisee for the company InstantWorld (a ficti-
tious company), which seeks to open the highest speed Internet café in
the City and County of Denver. Demographic analysis that the parent company conducted showed that its Internet cafés do best in the following
neighborhoods:

1. In a neighborhood where the percent of 18 to 21 year-olds is over 10%
   of the total population of the neighborhood.
2. Within one kilometer of a high school, university, or college.
3. On a busy street (with a TIGER Census Feature Class Code (CFCC2) of
   A1, A2, or A3).

3.5.4 Data management: Getting data sets and getting them ready for analysis

Create a folder on your computer or on the network where you will store the
data. Make certain that this folder has a name that is logical so that you will
understand what its contents are, and without spaces to avoid any problems
in your GIS-based analysis. Go to the following website to download the data:
http://www.esri.com/data/download/census2000-tigerline/index.html

You will see the TIGER 2000 Geography Network page, screenshot displayed
in **Figure 3.9**.

Figure 3.9 *Data source. Source of base map: Esri software.*

Once on the TIGER 2000 Geography Network page, select "Preview and Download" on the left side, and on the next screen, select Colorado via the map or via the list. On the next screen, select Denver County and press "Submit Selection." Do not select any layers yet. On the next screen, under "Available Data Layers," select the following five boxes, and then press "Proceed to Download" at the bottom of the list: Block Groups—2000—make sure you access block groups and not "Blocks"; Landmark Points; Landmark Polygons; Line Features—Roads; Census Block Group Demographics (under "Available Statewide Layers"). Scroll down as needed to find these layers. Select "Proceed to Download." Next, you will receive a notice that your data file is ready. Click Download File to save your data to the folder you created at the beginning of this activity.

Unzip the file with WinZip, 7-Zip, the extractor that comes with Windows XP/Vista/7, or other unzipping program. Note that there will be five files that are zipped underneath this "master level" zip file, plus a readme.html file. Be sure to Unzip each of the files and extract. When you are finished, you will have the following files: (1) tgr08000sf1grp.dbf contains the demographic data for Colorado; (2) tgr08031grp00.dbf, shp, shx contains the block group polygon boundaries for Denver; (3) tgr08031lkA.dbf, shp, shx contains the geometry for the roads; (4) tgr08031lpt.dbf, shp, shx contains the Census landmark points, such as schools and hospitals; (5) tgr08031lpy.dbf, shp, shx contains the Census landmark polygons; and (6) readme.html contains the metadata for the Census and GIS data that you will be using. The US Census Bureau collects data for political areas and statistical areas, and you will use some of this data for this project. A later chapter will go more into the detail of this remarkable source of data.

Look at the coordinate system statement on the location from which you downloaded the data: http://www.esri.com/data/download/census2000_tigerline/index.html. What is the coordinate system of the data? You will see that the coordinate system is straight latitude–longitude geographic. In other words, the data is not in any sort of map projection. Organizations will typically store spatial data in this way, so that it can be easily read by GIS software and projected into the map projection of the user's choice.

Start ArcGIS by accessing the software. Then, navigate to the ArcCatalog application, and navigate to your workspace. Access the metadata tab and look up the spatial metadata for each layer. Has a map projection/coordinate system been explicitly defined for these data sets? As you suspected from your analysis of the map projection in the metadata, no map projection has been defined for any of these data layers.

You would be fine with working statistically with the data as is, without a map projection, but since you want to work with it not only statistically but also spatially, including drawing buffers and other functions that require spatial analysis, you need to define the data's coordinate system according to what was indicated on the TIGER download site. Otherwise, a buffer on a point in the northern part of your data set would not have the same dimensions as a buffer in the southern part of your data set, due to the curvature of the Earth.

Launch the ArcMap application from ArcCatalog to launch ArcMap with a blank map document. Access ArcToolbox → Data Management Tools → Projections and Transformations → Define Projection, (or just use "search" to search and find the Define Projection tool) and define the projection on your tgr08031lpt. shp shapefile. This is your landmark points data set. Define the map projection as Geographic Coordinate System → World → WGS 84 (World Geodetic System is what WGS stands for). Do the same for the other layers. For the other layers, you can save time by using the Import function when you are defining the projection, to import the defined coordinates for the tgr08031lpt file and applying this projection to the other layers. With the map projected using WGS 84, buffers you calculate will be geodesic buffers. Buffers around points will be ellipses.

Add all of the layers (shape files) that you downloaded earlier to your data frame: Roads, block group boundaries, landmark points, and landmark polygons. Rename the appropriate layers to "roads," "block group boundaries," "landmark points," and "landmark polygons" as appropriate, so that the file names are more intuitive. Name your data frame "Denver County." Save your map document with a logical name, such as "Internetcafesites," and place it into the appropriate folder. Your map document is stored as an "mxd" file. What are your map units for your data? Your map units are in decimal degrees. What is the coordinate system of your spatial data? The coordinate system is geographic, latitude–longitude. Set your display units to kilometers. The display units are what will be displayed in your scale and when you do any sort of measurement on your map. Your data sets are now ready for use in analysis. **Figure 3.10** shows these data sets mapped in WGS 84 with display units set to kilometers.

3.5.5 Analyzing your data: Buffers

In this chapter, we focus on the part of this activity that involves building buffers to determine proximity to schools. The full activity, linked below, will guide you through other aspects of problem solution such as finding regions with dense groupings of potential users and nearness to busy roads. We will visit these again in this book in later chapters.

To find educational proximity zones, first find out where the high schools are located. Examine the table for your point landmark features. Census Feature Class Code (CFCC) of D43 indicates if a point landmark feature is a school.

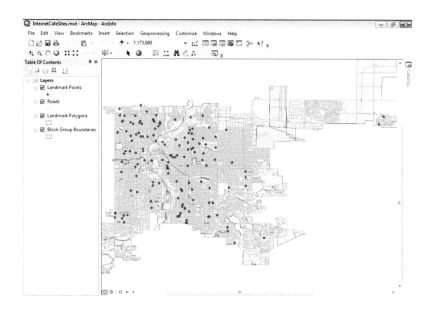

Figure 3.10 *Denver Internet café base map. Source of base map: Esri software.*

Select the schools in the point landmark feature layer. Recall that the Internet cafés do best within one kilometer of educational institutions. In ArcToolbox, use Analysis Tools → Proximity → Buffer to create one kilometer buffers around the selected features (schools). Because this layer is projected in WGS 84, the buffers will appear to be elliptical (with unequal major and minor axes). If they appear as circles (ellipses with equal major and minor axes) go back and check to make sure that the layer is properly projected (probably it is not).

Next, you need to find out where the colleges and universities are located. Examine the table for your polygon landmark feature. What CFCC indicates if a polygon landmark feature is a college or university? This is code "D43." As you did for the high schools, create one kilometer buffers for the colleges and universities. These buffers have an irregular shape because they are buffering irregularly-shaped areas.

Now you have created educational proximity zones. **Figure 3.11** shows the results of this analysis placed on the base map of **Figure 3.10**. Intersecting these buffers, later, with block groups containing at least 10% 18 to 21 year-olds will lead to a solution to the problem when coupled with nearness-to-busy-street variable and other elements of analysis. Multiple skill sets, and multiple layers of analysis, are often involved in analysis involving spatial mathematics!

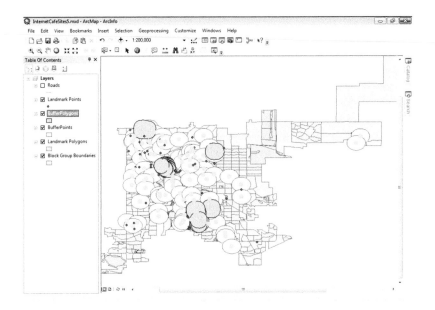

Figure 3.11 Buffers overlain on Denver Internet café base map. Yellow geodesic buffers are regularly shaped elliptical buffers of points. Orange geodesic buffers are irregularly shaped buffers of irregularly shaped areas. Source of base map: Esri software.

3.6 Related theory and practice: Access through QR codes

Theory

Persistent archive:

University of Michigan Library Deep Blue: http://deepblue.lib.umich.edu/handle/2027.42/58219

From Institute of Mathematical Geography site: http://www.imagenet.org/

Arlinghaus, S. L. and W. E. Arlinghaus. 2011. The Perimeter Project, Part 6. *Solstice: An Electronic Journal of Geography and Mathematics*. Volume 22, No. 2. Ann Arbor: Institute of Mathematical Geography. http://www.mylovedone.com/image/solstice/win11/Arlinghaus2011.pptx

Arlinghaus, S. L. 2001. Base Maps, Buffers, and Bisectors. *Solstice: An Electronic Journal of Geography and Mathematics*. Volume XII, No. 2. Ann Arbor: Institute of Mathematical Geography. http://www-personal.umich.edu/~copyrght/image/solstice/win01/sarhaus

Arlinghaus, S. L. and W. C. Arlinghaus. 2001. The Neglected Relation. *Solstice: An Electronic Journal of Geography and Mathematics.* Volume XII, No. 1. Ann Arbor: Institute of Mathematical Geography. http://www-personal.umich.edu/~copyrght/image/solstice/sum01/compplets.html

Arlinghaus, S. L., F. L. Goodman, D. A. Jacobs. 1997. Buffers and Duality. *Solstice: An Electronic Journal of Geography and Mathematics.* Vol. VIII, No. 2. Ann Arbor: Institute of Mathematical Geography. http://www-personal.umich.edu/~copyrght/image/solstice/win97/solsb297.html

Practice

From Esri site: http://edcommunity.esri.com/arclessons/arclessons.cfm

Kerski, J. Analyzing Data. Complexity level: 1, 2. http://edcommunity.esri.com/arclessons/lesson.cfm?id=493

Kerski, J. Siting An Internet Café in Denver Using GIS. Complexity level: 3. http://www.mylovedone.com/Kerski/Denver.pdf

Kerski, J. Analyzing Data. Complexity level: 3a. http://edcommunity.esri.com/arclessons/lesson.cfm?id=498

Kerski, J. Analyzing Data. Complexity level: 3b. http://edcommunity.esri.com/arclessons/lesson.cfm?id=504

Kerski, J. Analyzing Data, ArcGIS version. Complexity level: 3. http://edcommunity.esri.com/arclessons/lesson.cfm?id=524

Kerski, J. Asking Questions of Your Data 1. Complexity level: 3. http://edcommunity.esri.com/arclessons/lesson.cfm?id=505

Kerski, J. Asking Questions of Your Data 2. Complexity level: 3. http://edcommunity.esri.com/arclessons/lesson.cfm?id=506

Kerski, J. Traveling By Land and By Sea. Complexity level: 3. http://edcommunity.esri.com/arclessons/lesson.cfm?id=518

Replication of Results
Color and Number

Keywords: Color (94), Earth (33), map (31), image (30), divisible (17)

> I never saw a moor,
> I never saw the sea,
> Yet know I how the heather looks
> And what a wave must be.
>
> **Emily Dickinson**

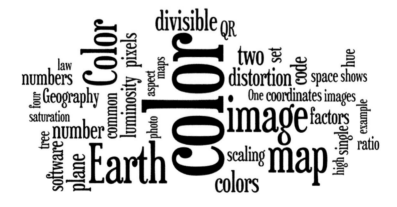

4.1 Introduction

The balance between art and science is a delicate one. A beautiful painting is often admired by all; but it is unique. No one else can replicate it. A major goal, however, in scientific research is to have the capability for independent practitioners to be able to repeat experiments so that the outcome is consistent. That goal is particularly difficult to achieve when the dynamic Earth is the laboratory. Establishing controlled experiments is hard. Imagination can help; visualization can help; and, art can help.

As we have seen in previous chapters, one way to ensure the replication of results is to have a clear system for organizing information. In the case of creating a replicable coordinate system, standard circular measure coordinates were used to transform the relative locational system of counting parallels and meridians to the absolute, and replicable, system of latitude and longitude. We have also seen, particularly in the "practice" sections, that organization of information in tabular form, as an attribute table in a GIS map, is a helpful and natural scheme. Clear thinking, coupled with clear organization, enables replication of results.

Beyond two organizational tools, one visual (a table, or a spreadsheet), and one numerical (the latitude/longitude system), there are tools that are more subtle that can be used for organizing geographic and mathematical information. In this chapter, we explore two of these subtle tools. Again, one is visual, color; and the other is numerical, prime factorization. We interpret both these concepts in terms of maps and conclude, in the practice section, with an exercise involving color and number.

4.2 Background—Color

Background is important not only in color visualization but also in fostering a deep understanding of a variety of abstract concepts. One place to begin any background study of color is with the four-color problem (now, "theorem"; Appel and Haken, 1976). For centuries, mathematicians have concerned themselves with how many colors are necessary and sufficient to portray maps of a variety of regions and themes. For example, two regions were said to be adjacent, and therefore required different colors, if and only if they share a common edge; a common vertex, alone, was not enough to force a new color. The answer of how many colors to use depends on the topological structure of the surface onto which the map is projected. When the map is on the surface of a torus (doughnut), seven colors are always enough (the reader interested in discovering the reasons why seven is considered "enough" is referred to the section on extra readings at the end of this book). Surprisingly, perhaps, the result was known on the torus well in advance of the result for the plane (then again, the plane is unbounded and the torus is not). The same number of colors that work for the plane will also work for the surface of a sphere (viewing the plane as the surface of the sphere with one point removed). How is the plane a spherical surface with one point removed? Through stereographic projection, everything on the spherical surface but the North Pole maps into a plane tangent to the sphere at the South Pole (more detail to come later in **Chapters 9 and 10**). It was not, however, until the last half of the twentieth century, aided by the capability of contemporary computing equipment to examine large numbers of cases, when the age-old "four color problem" became the "four color theorem." Appel and Haken (1976) showed that four colors are always enough to color any map in the plane. This discovery prompted the University of Illinois to issue a postage meter stamp of "four colors suffice" announcing this result of the solution of one of the world of mathematics' great unsolved problems.

The world of creating paper maps and publishing them has traditionally been one that is black and white: Color processing is expensive and often has been prohibitive. Nonetheless, cartographers, photographers, and others have developed a number of strategies for considering color, independent of how many colors suffice to color a map in the plane. Indeed, Arthur Robinson noted (Robinson, 1960, p. 228):

> Color is without a doubt the most complex single medium with which the cartographer works. The complications arise from a number of circumstances, the major one being that even yet we do not know precisely what color is. The complexity is due to the fact that, so far as the use of color is concerned, it exists only in the eye of the observer.

Like the mathematician, the cartographer, too, has significant unsolved problems associated with the concept of color.

Thus, color choice and use is typically tailored to "standard" reactions, by a typical observer, to color. The effect of color on an observer is often captured using the following terms as primitive terms: Hue, saturation, and luminosity.

- Hue is the term used to describe basic color. Blue, red, and green are all hues. White light passing through a prism is broken up into the spectrum of the rainbow composed of a variety of hues. The basic hues evident in this process are often referred to as spectral hues and these can be used to generate a progression of interstitial hues to fill in between the evident hues.
- Saturation is a measure of the amount of hue in a color; it is also referred to as intensity. Thus brilliant colors are more intense than are light pastels.
- Luminosity is a measure of relative lightness or darkness of a color. Color can be matched against a gray scale to make this measurement. One would expect, for example, that most shades of yellow are lighter or more luminous than most shades of red.

The contemporary computing environment of today, with its high-resolution color screens on the desktop computer and on hand-held devices including smartphones and data loggers, has fostered a whole new world of communicating data with color. Yet, an understanding of color theory is still essential. One could argue that the digital environment makes it more important than

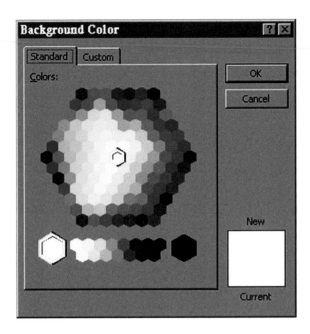

Figure 4.1 *Color wheel. Derived from Netscape 7.2. Source: Arlinghaus, S. L. and W. C. Arlinghaus. 1999. Animaps III: Color Straws, Color Voxels, and Color Ramps. Solstice: An Electronic Journal of Geography and Mathematics. Volume X. No. 1. Ann Arbor: Institute of Mathematical Geography. Source of base image: Netscape software. http://www-personal.umich.edu/%7Ecopyrght/image/solstice/sum99/animaps3.html.*

ever to understand how color can be used to communicate, particularly with a rich medium such as maps. On the desktop computer, users of various software packages in common use are exposed to the hue–saturation–luminosity set of primitive terms on a regular basis. In addition, they see the RGB (Red–Green–Blue) description using three primitive terms and the environment of their printer and photocopier's toner and layers based on CMYK (Cyan–Magenta–Yellow–Black). A color wheel can help the user to design strategies for color change: To decrease magenta, for example, subtract magenta, or add cyan and yellow (opposite from magenta) (**Figure 4.1**).

4.3 Color straws and color voxels

One obvious way to look at color, given two sets of primitives each with three elements, is as an ordered triple in Euclidean three-space (Arlinghaus and Arlinghaus, 1999). Indeed, that is how color maps are set up in older or contemporary software such as Netscape, Microsoft Office, Adobe PhotoShop, and so forth. Hue is measured across a horizontal x-axis (**Figure 4.2**) and saturation is measured along a vertical y-axis (**Figure 4.3**). The result is a square or rectangle with vertical strips of color corresponding in order to the

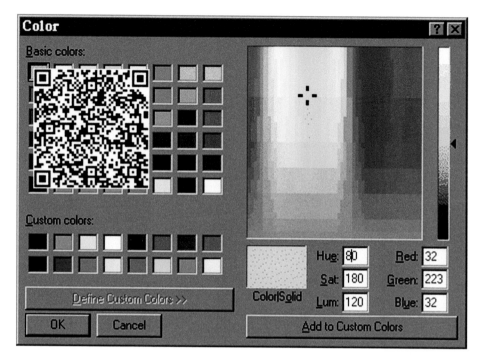

Figure 4.2 *Animated color map (linked to QR code): http://www-personal.umich. edu/~copyrght/image/solstice/sum99/hue.gif. Shows change in the resulting hue as one moves across the x-axis. Derived from Netscape 7.2. Source: Arlinghaus, S. L. and W. C. Arlinghaus. 1999. Animaps III: Color Straws, Color Voxels, and Color Ramps. Solstice: An Electronic Journal of Geography and Mathematics. Volume X. No. 1. Ann Arbor: Institute of Mathematical Geography. Source of base image: Netscape software. http://www-personal.umich.edu/%7Ecopyrght/image/solstice/sum99/animaps3.html.*

pattern on the color wheel. A third axis of luminosity (a gray scale) is often seen as a strip to the right of this square (**Figure 4.4**). It serves to match the selected color against light/dark values.

These animated color maps fix two dimensions and allow a third one to vary. That variation shows up in the small rectangle to the lower left of the color map and also in the "straw" to the right of the plane region. In all three cases, hue is the variable mapped on the horizontal axis, saturation is the variable mapped on the vertical axis, and luminosity is the variable mapped in the straw to the right. Once again, axes are important. Thus, in **Figure 4.2**, luminosity is fixed at 120 as indicated by the small arrow to the right of the straw. Saturation is fixed at 180 along the left side of the rectangle. Only hue is allowed to vary, as shown in the progression of the crosshair movement. The small rectangle to the lower left of the color map changes in color to show the hue of the current position of the crosshair. Thus, to see a hue-straw, one would need to take all 256 colors available in the flashing rectangle and stack them up in order of progression. Similarly,

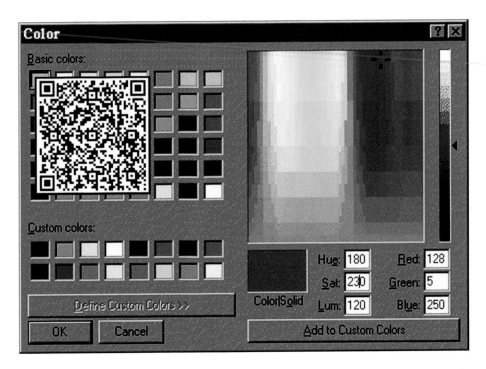

Figure 4.3 Animated color map (linked to QR code): http://www-personal.umich. edu/~copyrght/image/solstice/sum99/sat.gif. Shows change in the resulting saturation as one moves along the y-axis. Derived from Netscape 7.2. Source: Arlinghaus, S. L. and W. C. Arlinghaus. 1999. Animaps III: Color Straws, Color Voxels, and Color Ramps. Solstice: An Electronic Journal of Geography and Mathematics. Volume X. No. 1. Ann Arbor: Institute of Mathematical Geography. Source of base image: Netscape software. http://www-personal.umich.edu/%7Ecopyrght/image/solstice/sum99/animaps3.html.

one can allow saturation to vary and can keep hue and luminosity fixed (**Figure 4.3**). When luminosity is once again fixed at 120, and hue at 180, a structurally identical situation occurs (to that above). To see a saturation-straw, one would need to take all 256 colors available in the flashing rect-angle and stack them up in order of progression. The final case, in **Figure 4.4**, keeps hue and saturation fixed and allows luminosity to vary. Thus, one imagines a point in the base hue/luminosity plane fixed at (180, 120) and variable height shown in the luminosity straw reflecting changes in the single color-point as one alters luminosity. In this latter case, the obvious straw that appears is in fact the actual luminosity straw sought. In two of the cases, there is no evident straw of color and in the third there is; visualiza-tion is not impossible but it is made difficult.

An alternate way to visualize all this is to think of a cube (in three-space) of 256 units on a side. Label the *x*-axis as hue, the *y*-axis as saturation, and the *z*-axis as luminosity. Then, draw a plane parallel to the base plane (bottom

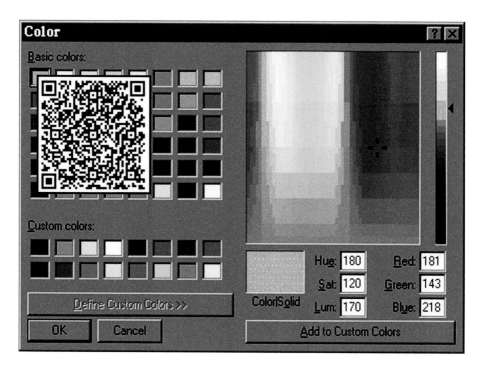

Figure 4.4 Animated color map (linked to QR code): http://www-personal.umich. edu/~copyrght/image/solstice/sum99/lum.gif. Shows change in the resulting luminosity as one moves along the z-axis. Derived from Netscape 7.2. Source: Arlinghaus, S. L. and W. C. Arlinghaus. 1999. Animaps III: Color Straws, Color Voxels, and Color Ramps. Solstice: An Electronic Journal of Geography and Mathematics. Volume X. No. 1. Ann Arbor: Institute of Mathematical Geography. Source of base image: Netscape software. http://www-personal.umich.edu/%7Ecopyrght/image/solstice/sum99/animaps3.html.

of the cube) at height 120 (we invite the reader to try it, using pencil and paper). Fix lines at 180 within that plane: One with hue = 180 and one with saturation = 180. These two lines trace the paths of the crosshairs, respectively, in **Figures 4.2 and 4.3**. What the cube approach also shows clearly is that there is really a set of voxels (volume pixels) making up the cube: There are 256 straws available for each of the three variables. Since $256 = 2^8$, there are therefore $2^8 * 2^8 * 2^8 = 2^{24} = 16,777,216$ voxels within the color cube (note the reliance on discrete mathematics and discrete structuring of a normally continuous object).

The notion of looking only at voxel subsets within a single plane parallel to a face of the cube is limiting within this large, but finite, set of possibilities. In choosing sequences of color, there may well be reason to follow a diagonal, to tip a plane, or to find various other ways of selecting subsets of color, as a smoothed color ramp, from this vast array. It is to these possibilities that we now turn.

4.4 Color ramps: Alternate metrics

The problem of finding color ramps linking one color to another can be captured simply as follows. To find a ramp joining two colors, A and B, first represent each of A and B as an ordered triple in color voxel space. Then, the problem becomes one of "find a path from A to B." Because one is limited to integer-only arithmetic, divisibility of distances often will not be precise; thus, one is thrown from the continuous realm of the Euclidean metric into considering the non-Euclidean realm of the Manhattan metric (of square pixel/cubic voxel space). The algorithms for finding shortest paths between two arbitrary points using integer-only arithmetic will therefore apply to colors mapped in color space as well as to physical locations mapped on city grids (as so-called Manhattan space). To see how these ideas might play out with colors, we consider an example that will lead to an animated color ramp.

Find a path through color voxel space from (80, 100, 120), shown in **Figure 4.5a** as a medium green to (200, 160, 60), shown as a fairly deep purple in **Figure 4.5b**. One set of points through which to pass, spaced evenly (not always possible), is given in the table below. The left-hand column shows the values of hue, the middle column shows the values of saturation, and the right-hand column shows the values of luminosity.

80	100	120
90	105	115
100	110	110
110	115	105
120	120	100
130	125	95
140	130	90
150	135	85
160	140	80
170	145	75
180	150	70
190	155	65
200	160	60

Figure 4.6 shows a screen capture and a QR code leading to an animation using the path outlined in the table above. The crosshairs show the movement along the path while the flashing color in the rectangle below the color map shows the associated color ramp. Clearly, the choice of path is not unique: Geodesics are not unique in Manhattan space. From this analysis, we see that the following theorem will hold.

Theorem: The determination of color ramps joining two colors is abstractly equivalent to finding paths in Manhattan space between two arbitrary points (where geodesics are not unique) (Arlinghaus and Arlinghaus, 1999).

(a)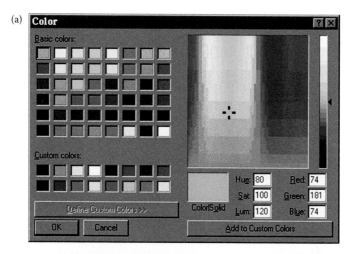

(b)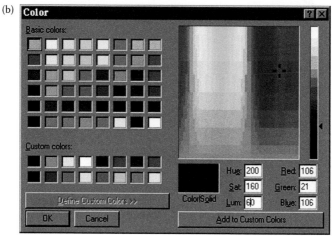

Figure 4.5 *(a) Beginning of a path through color voxel space. (b) End of a path through color voxel space. Derived from Netscape 7.2. Source: Arlinghaus, S. L. and W. C. Arlinghaus. 1999. Animaps III: Color Straws, Color Voxels, and Color Ramps. Solstice: An Electronic Journal of Geography and Mathematics. Volume X. No. 1. Ann Arbor: Institute of Mathematical Geography. Source of base image: Netscape software. http://www-personal.umich.edu/%7Ecopyrght/image/solstice/sum99/animaps3.html.*

We shall revisit the ideas embodied in this theorem in the last chapter of this work when we look briefly at routing problems and path-finding. The deeper related mathematical questions center on adjacency, graph theory, and topology. Follow the link to the four-color theorem at the end of this chapter; think about issues associated with map coloring in terms of adjacency of colored units on different surfaces (as noted earlier in this chapter).

Further, one might wonder what would happen with color in higher dimensions. In a related but not identical manner, the RGB scheme may also be

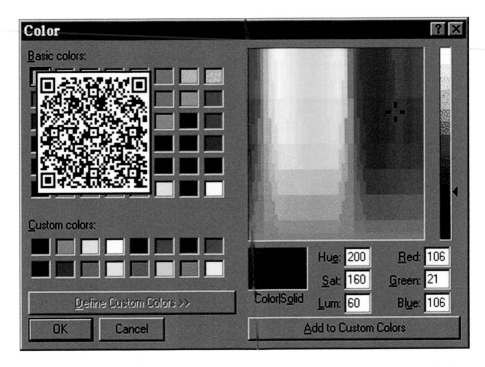

Figure 4.6 *Animation (linked to QR code) and screen capture of a color ramp leading from beginning in Figure 3.5a to end in Figure 3.5b. http://www-personal.umich. edu/%7Ecopyrght/image/solstice/sum99/ramp1.gif. Derived from Netscape 7.2. Source: Arlinghaus, S. L. and W. C. Arlinghaus. 1999. Animaps III: Color Straws, Color Voxels, and Color Ramps. Solstice: An Electronic Journal of Geography and Mathematics. Volume X. No. 1. Ann Arbor: Institute of Mathematical Geography. Source of base image: Netscape software. http://www-personal.umich. edu/%7Ecopyrght/image/solstice/sum99/animaps3.html.*

represented as describing color using three-space. In that scheme, the gray scale comes out as a 45 degree diagonal. Computer scientists offer a color code containing six alphanumeric characters, appearing in pairs of hexadecimal code that also serve as a three-space. Generally schemes such as this one, offer only visual slices through this three-dimensional (3D) color space along axes or in other "expected" ways. However, different vantage points offer different perspectives. Pantone color formula guide books offer one physical set of straws by which to probe 3D color space. The theorem above offers a comprehensive mathematical set.

4.5 Algebraic aspects of ratios

When one inserts an image into an electronic document, a set of handles appears on the image permitting the reader to stretch the image. If a single handle is simply pulled, distortion of the original may enter the picture. The result of attempting to enlarge or reduce the size of the image, by hand, does not permit replication of the original image. To be certain of such replication, one must preserve the aspect ratio of the image. Doing this usually requires the corner of the image to be pulled, to ensure that two sides will be pulled, rather than only one side. The aspect ratio is the ratio of the height to the width of the image. All modern software permits such preservation; sometimes as the default, sometimes as an option to choose, and sometimes by entering numbers of pixels or inches for each dimension.

The distortion of images can be significant in making terrestrial maps. If distortion is extreme enough, boundaries might not be recognizable. Navigation at sea or on land could fail, as was common for centuries even after sextants and other devices came into common use, resulting in the loss of ships, their cargo, and their sailors. With more unconventional maps, such as the QR code, distortion may completely ruin the map. The QR code in **Figure 4.7a** links to the website of the publisher of this work. When a small amount of distortion is introduced, as in **Figure 4.7b**, the code still links to the appropriate site. However, great distortion destroys the linkage (**Figure 4.7c**). It is a matter of contemporary research interest to discover the constraints within which QR code distortion is acceptable. The answer is not likely to be one set of constraints; it is likely to depend on why the distortion is occurring.

Indeed, why would a QR code become distorted? Other than the deliberate distortion in this document, or sloppy editing, or faulty use of word-processing software, it might seem that such a situation would not happen. However, when QR codes are transferred to other surfaces, distortion may occur. A lecturer with a laptop linked to a typical movie screen of low resolution may find

Figure 4.7 *(a) 100 pixels wide by 100 pixels high. Coded to connect to http://www. crcpress.com/. Connection should go through from a smartphone. (b) 100 pixels wide by 110 pixels high. Coded to connect to http://www.crcpress.com/. Connection should go through from a smartphone. (c) Clear distortion. Coded to connect to http://www.crcpress.com/. Connection might not go through from a smartphone.*

that QR codes become unreadable (through distortion). Or, the presentation of the QR code on plastic that is not dimensionally stable, as part of an advertising campaign, may cause distortion sufficient to inhibit communication. Or, the creation of cemetery QR code markers engraved in bronze may suffer, over the years, from the elements or from inconsistent engraving practices (Arlinghaus, W. E., 2011).

4.6 Pixel algebra

Number patterns underlie color, the scaling of images, and the effective use of color and other subtle cues to communicate spatial information. The algebra of pixels (Arlinghaus, 1996) is an integer algebra that can be viewed to operate on a bounded, finite set of pixels (the dimensions of the computer screen). In the image of **Figure 4.7a**, both dimensions are even numbers and so clearly both are divisible by 2. The aspect ratio, 100/100, can be preserved by taking 1/2 of each dimension—50/50. Are both the numerator and the denominator divisible by 3? Clearly not, so that while 1/2 is a scaling factor that works for this image, 1/3 would not be such a scaling factor.

Distributive Law. For integers in a domain of two operations, + and *: Suppose that a, b, and c are integers. Then a*(b + c) = a*b + a*c.

One application of the distributive law yields the trick that to determine whether or not a number is divisible by three, add the digits which form it, and repeat the process until a single digit is reached. If that single digit is divisible by 3, then the entire number is divisible by 3. Consider, for example, an arbitrary image of dimension 810 pixels by 922 pixels. The 810 dimension becomes $8 + 1 + 0 = 9$ which is divisible by 3 so that 810 is also divisible by 3. The 922 dimension becomes $9 + 2 + 2 = 13$, or $1 + 3 = 4$, which is not divisible by 3 so that 922 is also not divisible by 3. Thus, 1/3 would not be a scaling factor that would preserve the aspect ratio for the arbitrary figure of 810×922.

To see why this procedure is an application of the distributive law, note that numbers such as 9, 99, and 999 are always divisible by 3:

> $810 = 8*100 + 1*10 + 0*1$—first use of the distributive law
> $= 8*(99 + 1) + 1*(9 + 1) + 0*1$—partitioning of powers of 10 into convenient summands
> $= 8*99 + 8*1 + 1*9 + 1*1 + 0*1$—second use of the distributive law
> $= 8*99 + 1*9 + 8*1 + 1*1 + 0*1$—commutative   law   of   addition
> ($a + b = b + a$); note that the sum $8*99 + 1*9$ is divisible by 3 because 9 and 99 are divisible by 3
> $8*99 + 1*9 = (8*33 + 1*3)*3$—third use of the distributive law, so now,
> $810 = (8*33 + 1*3)*3 + (8*1 + 1*1 + 0*1)$ and for the number 810 to be divisible by 3 (fourth use of the distributive law)

all that is thus required is for the right summand in parentheses, 8*1 + 1*1 + 0*1 to be divisible by 3. Hence, the rule of three, for determining whether or not a given integer is divisible by 3 becomes clear.

A corresponding rule of nine is not difficult to understand, as are numerous other shortcuts for determining divisibility criteria. Clearly, one needs to test candidate divisors only up to the square root of the number in question. However, when one is faced with an image on the screen, it would be nice not to have taken the trouble (however little) of finding that 810 is divisible by 3, only to find that 922 is not divisible by 3. A far better approach is to rewrite each number using some systematic procedure and then compare a pair of expressions to determine, all at once, which numbers are divisors of *both* 810 and 922. For this purpose, the Fundamental Theorem of Arithmetic is critical.

Fundamental Theorem of Arithmetic (e.g., Gauss, 1801; Herstein, 1964): *Any positive integer can be expressed uniquely as a product of powers of prime numbers (numbers with no integral divisors other than themselves and 1).*

Thus, in the 810 by 922 example, $810 = 2*405 = 2*5*81 = 2*3*3*3*3*5$ (superscripts avoided by repetitive multiplication) and, $922 = 2*461$. The number 810 was easy to reduce to its unique factorization into powers of primes; one might not know whether or not 461 is a prime number or whether further reduction is required to achieve the prime power factorization of 922.

To this end, the Sieve of Eratosthenes (the same Librarian at Alexandria who measured the circumference of the Earth, described in **Chapter 2**) works well. To use the sieve, simply test the prime numbers less than the square root of the number in question. The square root of 461 is about 21.47. So, the only primes that can possibly be factors are: 2, 3, 5, 7, 11, 13, 17, and 19. Clearly, 2, 3, and 5 are not factors of 461. A minute or two with a calculator shows that 7, 11, 13, 17, and 19 are also not factors of 461.

Notice that these calculations need only be made once—when one has the unique factorization, all divisors of both numbers are known from looking at the two factorizations, together. Thus:

$$810 = 2*3*3*3*3*5; \quad 922 = 2*461$$

and 2 is the only factor common to both numbers, so that 1/2 is the only scaling factor for this image.

4.7 Preservation of the aspect ratio

Because the only factor the two numbers 810 and 922 have in common is 2, it follows that the only scaling factor that can be used, that will preserve the aspect ratio, is 1/2. Had 1/3 or some other ratio been employed, a distorted view of the original image would have been the result.

The level of sensitivity to image distortion will vary with the individual; **Figures 4.7a and b** may appear the same to some. **Figures 4.7a and c** will appear different to more. It is important to have absolute techniques that guarantee correct answers. Once one knows them, then one can choose when and when not to violate them.

The reduction of the original by 1/2, producing an image 1/4 of the original size, may still not be desirable. One can use the unique factorization to build what is desired. The value of 810 has a number of factors; thus, one might choose to rescan the image, holding the width at 810 pixels and shaving just a bit off the height. Now, to create the possibility of various scaling factors, consider a tiny sliver removed to create a height of 920 = 2*2*2*5*23; there is an extra factor of 5 that 920 has in common with 810 that 922 did not, so all of 1/2, 1/5, and 1/10 (1 over 2*5) are scaling factors that will preserve the aspect ratio. However, one might wish for still more possible scaling factors; if a slightly larger sliver can be removed, so that the height of the scanned image is 900 pixels, with 900 = 2*2*3*3*5*5, then 810 and 900 have common prime power factors of 2, 3, 3*3, and 5, so that the set of scaling factors has been substantially expanded to include all of:

1/2, 1/3, 1/9, 1/5,
combination of four things one at a time—4!/(1!*3!)
1/6, 1/18, 1/10; 1/27, 1/15; 1/45
combination of four things two at a time—4!/(2!*2!)
1/54, 1/30, 1/90, 1/135
combination of four things three at a time—4!/(3!*1!)
1/270
combination of four things four at a time—4!/(4!*0!)

This set of values for the height, in pixels, should offer a variety of choices for scaling factors.

4.8 Image security

In creating images, it is useful to have an aspect ratio with the numerator and the denominator having many common factors; greater flexibility in changing the image is a consequence, as above. However, there may be situations, particularly on the Internet where downloading is just a right-click of the mouse away, in which one wishes to inhibit easy rescaling of an image. When that is the case, unique factorization via the Fundamental Theorem of Arithmetic again yields an answer: Choose to scan the image so that the numerator and the denominator of the aspect ratio are relatively prime—that is, so that the numerator and the denominator have no factors in common other than 1. In the case of the 810 by 922 image, altering the aspect ratio from the original of 810/922 to 810/923 would serve the purpose: 923 is not divisible by any of

2, 3, or 5. Anyone downloading the original could not resize it without distortion: Distortion, for once, becomes the mapmaker's friend.

4.9 Theory finale

The abstract lessons learned in modern algebra are as important in the electronic world as they are elsewhere. Indeed, the current technical realm therefore offers a refreshing host of new examples to motivate theory, as suggested in this simple scaling example that drew on a variety of algebraic concepts including: The distributive law, the commutative law of addition, the Fundamental Theorem of Arithmetic and the needed associated concept of prime number, facts involving square roots and divisors, the Sieve of Eratosthenes, basic material on permutations and combinations together with the convention that $0! = 1$, and the idea of relatively prime numbers. The web is a rich source of example when one brings a rich source of abstract liberal arts training to it. The problem of "replication of results" has a long and deep history, extending its roots into often surprising corners of the world of spatial mathematics.

In the following section, you will have the opportunity to apply the theory and discussion in this chapter. First, you will use a web-based GIS called ArcGIS Online to change the color and size of symbols on maps that you create, assessing as you do so how meaning and interpretation are altered in the process.

4.10 Practice using selected concepts from this chapter

4.10.1 Changing symbol color and size to enhance meaning on maps

Access the web-based GIS known as ArcGIS Online using a web browser: http://www.arcgis.com/home. In the contents search box, search for a map using the following phrase: "Extreme temperature same owner:jjkerski." On the results screen that appears upon searching for this phrase, click under the thumbnail to open the map in the ArcGIS.com map viewer. Alternatively, open a web browser and go directly to the map at the following URL: http://bit.ly/ NPviC0. Click on the show map legend button and you will see a map similar to that below (**Figure 4.8**).

This map shows the daily high and low temperature for each day of January and July 2011 in the USA. Toggle off the high and low temperatures for each of the four map layers. What pattern do you notice for the high and low temperatures, and what is the reason for these patterns? How does altitude, latitude, and proximity to oceans relate to the extreme temperatures, and why? How does the fact that each extreme temperature is drawn with the same symbol affect your understanding of the data? You should notice that the high temperatures migrate from the coastal areas to the interior of the country

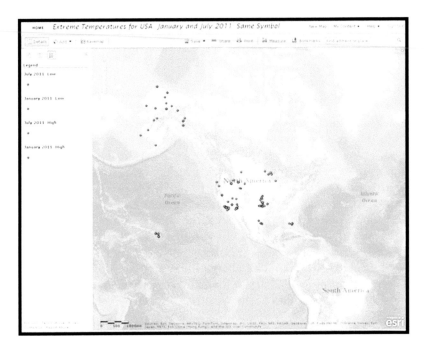

Figure 4.8 *Map showing the location of each city and state that recorded either the highest temperature in the country, or the lowest temperature in the country, for the months of January and July 2011. Source of base map: Esri software.*

from January through July, due to the moderating effect of the oceans. The January high temperatures also show a spatial relationship to latitude, as they all lie in Florida, southern Texas, southern California, and Hawaii. The effect of altitude, on the other hand, is to lower the temperature, as evident in the numerous lows in January and in July in Colorado, Montana, and other mountainous areas. The fact that the map is drawn using a single symbol is a hindrance to interpretation. Although toggling each layer on and off aids in interpretation, changing the symbology to a different symbol for each month of the year to differentiate between the highs and the lows, and changing the color and size during any particular month for a classification by temperature would all aid in understanding the data.

Click on cities until you find one that recorded more than one instance of an extreme temperature for the month. Does the fact that these cities are only symbolized once affect your interpretation of the most common locations for the extreme temperatures? In a similar way, select cities until you find one that was tied with another city for the high or low temperature for that day. Should those cities containing more than one extreme, or those that tied, be symbolized differently from the others? If so, how? These cities might be better symbolized as a larger or darker symbol to indicate that a tie or multiple records have occurred.

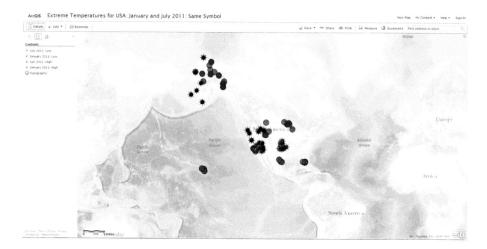

Figure 4.9 *Changing the color reveals extra information. Source of base map: Esri software.*

Select the "show contents of the map" button and change symbols for the January 2011 high temperatures. Change all of the symbols using "single symbol" to a red circle. Change the July 2011 high temperatures to a red circle with protruding spikes that looks like a red sun with rays. Change the January 2011 low temperatures to a blue circle, and the July 2011 low temperatures to a blue sun with rays (**Figure 4.9**). Can you more easily determine the changes between January highs and lows in this way? Can you determine the changes between January and July highs in this way? Does blue convey "cold" or "hot" to you? Does red convey "cold" or "hot" to you? Why? You should be better able to distinguish lows from highs, and January from July, with this symbology. Most often, red is used to convey heat, or "more" of something, while blue is used to convey "cold" or "less" of something, but this is simply map convention tradition. These traditions vary around the world and with time. For example, blue has traditionally been used to symbolize the oceans on European and North American maps, but green was traditionally used to symbolize the oceans on maps produced in Asia. If you have time, experiment with other symbology options, keeping in mind the color theory discussion from this chapter.

Edit the symbology for January 2011 again. This time, instead of "single symbol", select "color" and vary the color by temperature. Does this help or hinder your map interpretation? Color can effectively be used to distinguish between data types and data values. Repeat for the other three map layers. At which point are there too many color symbol types on the map for you to correctly interpret it? This is a fine line, and varies depending on such things as the data user, the data types, and the map scale, but at some point as the number of colors and symbols increase, the human brain cannot effectively distinguish between them. What, in your view, is the ideal color, symbol, and size to use for each map layer? What, in your view and in keeping with what you learned in this chapter, is the best color and symbol combination for the most effective map interpretation? The underlying base map affects the interpretation of the map. Experiment with changing the base map and be sure to try a light-gray canvas. What is the ideal base map to use for these data? It could be, in your view, the light-gray canvas, or perhaps the lightly symbolized National Geographic base map. Or, because you are examining the effect of altitude and landforms on temperature, a satellite image base map, even though it is not the simplest base map, might be the most effective.

4.10.2 Identifying and mapping trees for a stream bank erosion control project

Color and mapping can also greatly enhance the understanding of local issues and can aid in planning. Recently, Washtenaw County, Michigan has been embarking on a major stream bank erosion control project. When that project entered heavily forested residential lands adjacent to a creek, environmentally sensitive residents quite naturally became concerned for the trees and wildlife that would be destroyed or disturbed.

The County coded its easement with pink flags. It tagged selected large trees or otherwise interesting vegetation with a blue band if they were to be removed; it tagged trees within the easement with a red band if they were to be left alone. All vegetation within the easement, except trees or shrubs carrying red tags, was to be removed. Color was critical—a simple red/blue confusion could cost a tree its life!

One neighborhood used Google Earth, together with a GPS-enabled smartphone, to make an inventory of trees present, along a half-mile stretch of the creek, before the project began. David E. Arlinghaus did all the photography with a smartphone that geotagged the images. He then transmitted the images to S. Arlinghaus who did the mapping (**Figure 4.10**).

The accuracy of the geotagging of the photos was limited by several factors. First, the software in the smartphone has limits, as does the accuracy of the GPS signals in heavily wooded areas, as we have seen earlier in this book. Second, the geotagging of the photo is actually the geotagging of where he stood to take the picture of the tree, rather than of the tree itself. He attempted to stand at a consistent distance from trees to ensure accuracy (but that is difficult in a densely wooded area). The level of precision, however, was quite good—trees were in the correct position in relation to each other and in relation to the dwelling units.

The geotagged camera images were downloaded directly to a computer by plugging the smartphone into a new Windows 7 desktop computer. All 81 images were stored in a single folder. That folder was then uploaded to the free software called "GeoSetter." From there, the geotagged images were batch uploaded to Google Earth in a single operation (rather than entering each one individually). The GeoSetter software was able to take the underlying geocoded coordinates

Figure 4.10 *Pink arrows mark flags showing the County drain easements. Red balloons mark trees within the easement to be saved. Blue balloons mark good trees that will be cut. Source of base map and data: Google Earth mapping service © 2012 Google and Image US Dept of State Geographer, © 2012 Google, Data SIO NOAA US Navy NGA GEBCO, ©2012 MapLink/TeleAtlas.*

from the camera images, as well as the images themselves, and make them correspond to the underlying coordinate geometry in Google Earth. We made color decisions to correspond with the actual colors of tags used on the vegetation.

Accuracy of the registration of photo and Google Earth coordinates, using this sort of strategy, was guaranteed. Hand placement would not offer that level of accuracy of registration. Overall, the results were sufficiently precise (although not necessarily positionally accurate) to offer local residents a clear picture of what was going to happen in their wooded areas. When the camera GPS coordinates were obtained, a photo of the tagged item was also taken. **Figure 4.11** shows a photo displayed on the Google Earth surface pointing to the identified tree.

Figure 4.12 shows a similar configuration of a photo in relation to the Google Earth base. This time, the photo shows a tree with a blue tag. Note that there is a pointer on the margin of the photo that points to a blue balloon.

The neighborhood association established a tree monitoring committee. The committee was armed with a Google Earth file showing tree locations and associated tag colors. The easement was also geocoded. Prior to using the file, the neighborhood association president and the creator of the Google Earth display met with the lead County official on the project and the lead engineer

Figure 4.11 Photo mounted in Google Earth. Note that the photo has a pointer on it that points to the correct balloon location. Source of base map and data: Google Earth mapping service © 2012 Google and Image US Dept of State Geographer, © 2012 Google, Data SIO NOAA US Navy NGA GEBCO, ©2012 MapLink/TeleAtlas.

on the project to ensure agreement. Neighborhood involvement and monitoring was critical in developing a constructive relationship among the various parties adjacent to this project.

Next, try your hand at simple insertion of a balloon and an image directly into Google Earth. A County GIS file of streams overlain on Google Earth would show the creek from above as it passed through the dense woods. Load the free Google Earth on your computer or alternatively, use the Google Earth 3D "Earth Viewer" plug-in to your web browser. Zoom in to this area near N 42.2604 and W 83.6964. In Google Earth, pull down the "Add" pull-down menu (**Figure 4.13**). Choose "Placemark." Slide the placemark around and try to position it at N 42° 15′ 38.05″, W 83° 41′ 45.74″, the position of the tree in **Figure 4.12**. How confident are you that you can get the placemark in exactly the spot that was geotagged by the camera? How would you like to try this process for 81 different points? How confident would you be in the results? It is difficult, by hand, to position the mouse cursor at a precise latitude–longitude coordinate for one point, much less for numerous points.

Figure 4.12 Blue tagged tree. Source of base map and data: Google Earth mapping service © 2012 Google and Image US Dept of State Geographer, © 2012 Google, Data SIO NOAA US Navy NGA GEBCO, ©2012 MapLink/TeleAtlas.

One way to improve accuracy, directly in Google Earth, is to type in the coordinates (take care to make sure they are in a numerical format suited to the settings in Google Earth) in the "Fly to" slot under the "Search" arrow in the upper left-hand corner. Then press the magnifying glass button. The globe image will move to become centered on the coordinates typed in. Now, without moving the image, pull down the "Add" menu and choose "Placemark." The default setting is to locate the placemark in the center of the image so now the numbers typed in will match accurately with the placemark location. This strategy is both quicker and better than the one above. Still, though, there is no guarantee that the coordinates typed in are the best available coordinates. The camera may have stored them to more decimal places than we know about and so there may be error due to round off or truncation. As we have seen earlier, shaving off even a single number can cause difficulties in local studies. Now you should have greater appreciation for why we used extra software (GeoSetter, in this case) to ensure accurate registration of camera coordinates with Google Earth coordinates!

Figure 4.13 *Add a Placemark in Google Earth. Google Earth mapping service © 2012 Google and Image US Dept of State Geographer, © 2012 Google, Data SIO NOAA US Navy NGA GEBCO, ©2012 MapLink/TeleAtlas.*

Image insertion was also important to verify the tree type and its appearance at each tagged location. Try uploading a photo, by hand, into Google Earth. Go to the "Add" pulldown menu. Choose "Image Overlay" and upload any image you have on your computer. Again, try to position the image to imitate the result obtained in **Figure 4.12**. If you closed the image before you were done editing it, right-click on the name of the image overlay in the left-side panel, and choose "properties" to activate layer editing capabilities. As with inserting placemarks, the direct by-hand process is easy. Its accuracy, however, is limited. Again, improvement can be realized directly in Google Earth by typing coordinates in the "Fly to" box, as with the placemarks. Still, there are limits that are overcome easily using extra software to upload materials into Google Earth where we can take advantage of its interactive features. GeoSetter was easy and useful for this particular project. Esri GIS maps may also be brought directly into Google Earth and overlain on the imagery (as we did with the stream layers from the County). Further, Microsoft Excel spreadsheets may also be mapped in Google Earth, Google Maps, and in ArcGIS Online, as long as there are geocodable data columns within the spreadsheet. Depending on the software used, geocodable data columns include latitude–longitude or other coordinate system, or street addresses. Spreadsheet Mapper 2.0 v3 is available from Google Earth Outreach as a free download http://www.google.com/earth/outreach/tutorials/spreadsheet.html.

The important distinction here, between display software such as Google Earth and Google Maps versus GIS software, is that there is not an underlying attribute table in the display software in the same way that there is in the GIS software. Workarounds of various sorts are available that enhance the data interaction with the outstanding interactive display capability present in the Google Earth world. To make good decisions about which ones to choose, knowledge of how a GIS works is important.

4.11 Related theory and practice: Access through QR codes

Theory

Persistent archive:

University of Michigan Library Deep Blue: http://deepblue.lib.umich.edu/handle/2027.42/58219

From Institute of Mathematical Geography site: http://www.imagenet.org/

Arlinghaus, D. E. and S. L. Arlinghaus. 2012. Geosocial Networking: A Case from Ann Arbor, Michigan. *Solstice: An Electronic Journal of Geography and Mathematics.* Volume XXIII, No. 1. Ann Arbor: Institute of Mathematical Geography. http://www.mylovedone.com/image/solstice/sum12/Geosocial.html

Arlinghaus, S. L. and W. C. Arlinghaus. 1999. Animaps III: Color Straws, Color Voxels, and Color Ramps. *Solstice: An Electronic Journal of Geography and Mathematics.* Volume X. No. 1. Ann Arbor: Institute of Mathematical Geography. http://www-personal.umich.edu/~copyrght/image/solstice/sum99/animaps3.html

Arlinghaus, S. L. 1998. Animated Four Color Theorem: Sample Map. *Solstice: An Electronic Journal of Geography and Mathematics.* Volume IX, No. 2. Ann Arbor: Institute of Mathematical Geography. http://www-personal.umich.edu/~copyrght/image/solstice/win98/4color.html

Arlinghaus, S. L. 1996. Algebraic Aspects of Ratios. *Solstice: An Electronic Journal of Geography and Mathematics.* Volume VII, No. 1. Ann Arbor, Institute of Mathematical Geography. http://www-personal.umich.edu/~copyrght/image/solstice/sols196.html

Practice

From Esri site: http://edcommunity.esri.com/arclessons/arclessons.cfm

Examining the Tuscaloosa April 2011 Tornado in ArcGIS Online

Joseph J. Kerski

Parts 1 to 7 are linked below. Videos that appear on the YouTube Geographyuberalles channel. Look at the use of color.

Part 1: http://www.youtube.com/watch?v=oGiMKA5w2WM

Part 2: http://www.youtube.com/watch?v=GtGxTFhqgQ0

Part 3: http://www.youtube.com/watch?v=kSs1M833OSM

Part 4: http://www.youtube.com/watch?v=RroGnw0Yr1Y

Part 5: http://www.youtube.com/watch?v=BzDD9FEczTo

Part 6: http://www.youtube.com/watch?v=3UFp8C65ZHY

Part 7: http://www.youtube.com/watch?v=3UFp8C65ZHY

5

Scale

Keywords: Map (89), scale (66), dot (47), density (28), area (21)

> I have a little shadow that goes in and out with me,
> And what can be the use of him is more than I can see.
>
> …
>
> The funniest thing about him is the way he likes to grow—
> Not at all like proper children, which is always very slow;
> For he sometimes shoots up taller like an India-rubber ball,
> And he sometimes gets so little that there's none of him at all.

Robert Louis Stevenson

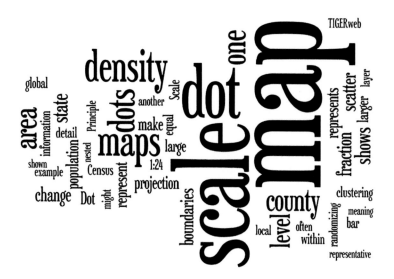

5.1 Introduction

The scale of a map represents the ratio of the distance between two points on a map to the corresponding distance on the surface of the Earth. Map Scale is Map Distance divided by Ground Distance. Suppose one inch on the map represents one mile on the surface of the Earth. There are a several common ways to express this idea. Verbal Scale simply states the relationship as: "One inch represents one mile." A Bar Scale shows graphically the relationship between the map distance and the surface distance, and is typically used to visualize distances on a map. Finally, map scale may be represented as a numerical ratio, called a Representative Fraction (RF), which is free from units. If, for example, RF = 1/24,000, then one meter on the map represents 24,000 meters on the surface of the Earth. Or, one foot on the map represents 24,000 feet on the surface of the Earth. It is convenient to capture the idea using notation such as 1:24,000, 1:63,360 (which is a scale of one inch to one mile), or 1:1,000,000.

Consider the difference in the following maps: **Figure 5.1a** (RF 1:24,000) shows a view of part of Chesapeake Bay; **Figure 5.1b** (RF 1:100,000) shows more of the bay, and **Figure 5.1c** (RF 1:250,000) shows more yet. Topographic maps of the United States are most often produced, by the United States

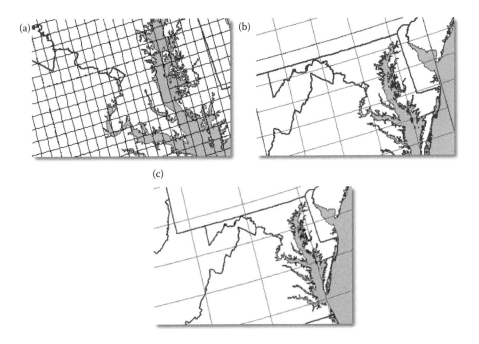

Figure 5.1 *(a) 1:24,000. (b) 1:100,000. (c) 1:250,000. Scale change displayed. Each cell depicts the area shown by maps at each scale indicated. The larger representative fractions display "large scale" maps. Source: National Atlas of the United States, March 5, 2003, http://nationalatlas.gov.*

Geological Survey (USGS), at these three scales. Maps in the 1:24,000 series depict an area that is 7.5 minutes of latitude by 7.5 minutes of longitude. In the 1:100,000 series, standard map size is 30 minutes of latitude by 60 minutes of longitude. In the 1:250,000 series, the standard map size is one degree of latitude by two degrees of longitude. The grid shown in **Figures 5.1a–c** depict the amount of area covered by maps at each of the three scales. The amount of territory depicted on each of the maps indicated by each "grid cell" is different because the scales are different. The shape of each of the map "grid cells" appears rectangular but it is actually an isosceles trapezoid because as you recall from our earlier discussion in this book, meridians converge at the poles. Scale, and geometry, matter!

5.2 Scale change

The Representative Fraction is of particular importance in showing changes in scale (**Figure 5.1**). Because the fraction 1/24,000 is larger than the fraction 1/50,000 (cut two equally sized pies into pieces—the pieces are smaller when 50,000 are cut than when 24,000 are cut), the map of scale 1:24,000 is said to have larger scale than the map of 1:50,000. Often, maps are called "large scale" maps. But, to give meaning to that phrase, there must be another map to which it is being compared. Certainly, the 1:24,000-scale map is the largest scale map in the set of three maps in **Figure 5.1**. Clearly, the 1:250,000 map is the smallest scale map in this set because the fraction 1/250,000 is smaller than either of the other two fractions. Is the map with a scale of 1:50,000 a large or a small scale map? The answer is that it depends on the context. In relation to the 1:24,000-scale map, it is smaller scale. But, in relation to the 1:250,000-scale map, it is larger scale.

Typically, topographic maps produced at 1:24,000 show natural features including rivers, valleys, and vegetation as well as constructed features such as railroads, roads, and buildings. The contour lines show the elevation above mean sea level. Maps at this scale are often used for urban planning projects, environmental management, and a variety of outdoor activities. Both the 1:100,000 series and the 1:250,000 series, no longer updated by USGS, show features similar to those shown in the 1:24,000 series but not in as much detail. Try your hand at creating maps at a variety of scales, online, using the MapMaker tool in the National Atlas. Zoom in to enlarge the scale and zoom out to reduce the scale. As we saw above, the terms "large" and "small" scale are relative terms and have meaning only when there are at least two maps considered in the discussion. That is because the reference is to comparison of representative fractions. Often, however, folks simply look at a map and say that because it shows detail it is a large scale map, or that because it shows a broad area without a lot of detail that it is a small scale map that is "generalized." In reality, however, *all* maps are generalized representations of the Earth, even those that show great detail. A map is only a representation of reality. Thus, it is often convenient, for clarity in communication, to call a map such as one that shows only a downtown area as a map displayed at "local" scale and one that shows a large part of a nation as a map displayed at "global" scale. Again, "local" and "global" scales are relative terms; they offer an alternative way to capture the parallel idea as "large" and "small" scale. Remember: A single map cannot be correctly characterized as being "large" scale—that is a relative term and a second map must be present for such a phrase to have meaning.

While the representative fraction is important in determining the size of the scale of one map in relation to another, the bar scale has become of increasing importance with the advance of technology. **Figure 5.2** shows a map of Michigan, produced using the National Atlas' MapMaker tool, with a bar scale in the lower right corner.

When a map is digitally or otherwise enlarged or reduced, its scale changes. But, if what is written on the original map is simply 1:50,000, then when the map is reduced, the representative fraction should change—one inch will no longer represent 50,000 inches (in fact, it will represent more than 50,000 inches). But, the scale stated on the reduced map will still read, 1:50,000 in reduced font size. The reduced map scale will be wrong. If, however, a bar scale appears on the original map, then that bar scale will also be reduced and it will correctly portray the scale of the new map. Since a tool such as the

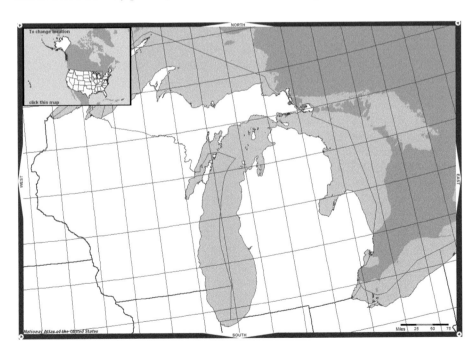

Figure 5.2 *Map showing bar scale. Source: Mapmaker, National Atlas of the United States, March 5, 2003, http://nationalatlas.gov.*

National Atlas Mapmaker permits users to change scale, it is appropriate that it employs bar scale to change in response to scale change.

This observation involving bar scale in relation to map scale change often becomes important in municipal planning applications. An urban planner may make a fine map on a huge piece of paper showing plans for a new shopping mall, complete with new tree location, parking spaces, onsite water retention plans, and so forth. But, unless that large original contains a bar scale, it will be wrong when it is reduced in size to present to the municipal planning commission for consideration. Municipal decision makers need to have accurate facts, including accurate maps—based on mathematics, in order to make informed decisions.

5.3 The dot density map: Theory and example

Probably we have all seen maps in the media that portray distributions as dots in geographic space. Often, these are simply ways to present material that might otherwise be captured by putting a wall map on a bulletin board and inserting push-pins at points of interest. In this model, the accuracy of pin placement determines whether the map is "right" or "wrong." Too often, though, this sort of bulletin board model is confused with a powerful, but perhaps underappreciated, style of map that shows clustering and relies on change of scale to do so. The latter is called a dot density map; we illustrate it, in theory and practice, below.

5.3.1 Construction of a dot density map

Figure 5.3 shows the beginning of a dot density map designed to represent the clustering of population in southeast Michigan. In it, the dots do not reflect single homeowner locations. A single dot represents some number of people living within a given county. The county boundaries nest inside the state boundaries; the county boundaries at the edge match perfectly with the state boundaries. The dot scatter is assigned to random locations within a base of polygons (in this case, counties). The bulletin board model described above is "accurate"; the dot density map is "precise." Mistakes in reading dot density maps come about by trying to assign "accuracy" where there may be little or none, such as in assuming that dot coordinates on the map represent accurate real-world locations at the latitude–longitude level rather than simply at the county level.

The concept of clustering is tied to scale and to scale change. The dots in a dot density map are simply a way to partition data. If the population of a county is 1,000,000 and a single dot represents a thousand people, then 1000 dots scattered randomly in that county might be used to represent that county's population. Viewed on its own, the dot scatter within a county boundary map such as this has little meaning (**Figure 5.3**). The county layer is being used

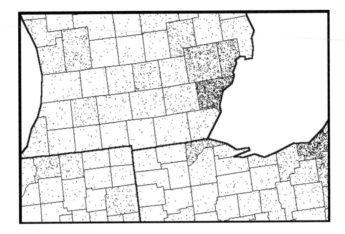

Figure 5.3 Dot scatter at the county level. The individual dots do not correspond to addresses or other geographic features. The scatter is assigned to random positions within a base of polygons—counties here. Source: Arlinghaus, S. L., unpublished lecture notes. Source of base map: Esri software.

Figure 5.4 Dot scatter viewed through the lens of state boundaries shows patterns of population clustering (with randomization of scatter taking place at the county level). Source: Arlinghaus, S. L., unpublished lecture notes. Source of base map: Esri software.

as a "randomizing layer." However, when the scale of viewing is changed, and when the county boundaries are removed and the state boundaries are shown, one gets a good picture of the clustering of population at the state scale through a state, rather than a county, lens (**Figure 5.4**). Further scale change may reveal broader patterns, as well. When looking at maps that purport to be dot density maps, it is critical to know if one is looking at the randomizing layer only or if the scale of the view has been altered so that clustering patterns are seen from a scale perspective broader than that afforded by only the randomizing layer.

In addition, there are a number of other subtle, but important, points that underlie the construction of a dot density map. Certainly we had to decide how many people were represented by a single dot. But, in addition, we have the option to choose whether such representation is absolute or relative. Absolute representation is one dot represents 1000 people. Relative representation is one dot represents 0.1% of the population of the state, county, or other areal unit. For the sake of communicating example in the sequence of images above, we chose an absolute representation. If instead, we had wished to make comparisons between regions by tossing a unit square on a map and comparing populations within selected unit squares to the random one, then choosing a relative representation would have been appropriate.

Finally, in order to make valid comparisons from one area to the next, make sure to choose an equal area projection (so that Greenland does not look bigger than Brazil, or another similar areal inaccuracy). This is another important, yet perhaps subtle, point that means more at the global level than it does at the local level. The base map chosen for the figure sequence above was an Albers Equal Area conic projection (more on projections in **Chapter 9**). When looking at a dot density map, consider therefore the projection on which it is based—it should be an equal area projection. Otherwise, areal distortions may lead to inaccurate conclusions about the density of the phenomenon being studied, because the dots may be spread out or clustered simply to cover the distorted areas.

To summarize the decisions:

- Select an equal area projection (such as an Albers Equal Area Conic for the United States).
- Select a distribution that can usefully be represented, in an absolute or relative manner, as dot scatter—such as population.
- Then, choose polygonal nets of at least two different scales—such as state and county boundaries. To extract the most meaning, make sure that the smaller units "nest" inside the larger units (rather than overfitting or underfitting them).
- Let the map with the smallest spatial unit that is practical (such as counties) be used as the randomizing layer—the dot scatter is spread around randomly within each unit.
- View the scatter through polygons (states) that are larger than are those of the randomizing layer (counties). **Figure 5.4** shows the results of removing the county boundaries. The clustering of dots at the state level means something; at the county level it is merely random.
- Since the underlying projection is an equal area projection, a unit square (or other polygon) may be placed anywhere on the map and comparisons may be made between one location and another.

5.3.2 Dot density map theory

Both analysis and synthesis play important roles in mapping. We use the dot density map to elicit validity-of-construction principles. The cartographic example presented below displays principles of spatial synthesis as they focus on centrality and hierarchy (Arlinghaus, 2005).

- Classical example: The dot density map employs a nested hierarchy of regions to convert information about dots to information about regions; in so doing, the clusters of dots emerge as centers of activity associated with the nature of the underlying data from which the dots were extracted.
- Contemporary example: The interactive online map may employ a nested hierarchy that, in a single map, offers not only information of the sort available in a dot density map but also a host of other previously impossible features, as well. It may be linked to the underlying database in a manner that also permits:
  - Scale transformation.
  - Views of the database corresponding to small regions on the map.
  - Searches of the underlying database.

Interactive capability can be far more than an interesting visualization tool; it can be one offering synthesis of spatial information at a level far greater than that available with any classical map.

A scatter of dots might represent any real world phenomenon, from the location of emergency telephone kiosks, to small villages, to national capitals. Dots pinpoint geographical position. At a global scale, national capitals may appear as dots; at a local scale, they may appear as areas containing dot scatter of their own. What matters is geographic scale: Dots at one scale may not be dots at another scale. In that regard, it does not matter what the dots represent. When the dot density map is properly constructed, it can serve as a tool offering valuable insight into clustering.

The pattern of boundary removal, to make sense of dot scatter, is not symmetric. In **Figure 5.3**, dots were randomized at the county level and then viewed at the state level. Since states are larger than counties and counties are nested within states, the opportunity to observe clusters at the state level is optimized. If the situation exhibited in **Figures 5.3 and 5.4** were reversed, and dot scatter were randomized at the state level, and then viewed through the county boundary lens, clear clustering errors would result.

Principle 1: Randomizing principle
In a dot density map, dot scatter is randomized at one scale and then, to have the map make sense, must be viewed at a scale more global than that of the randomizing layer.

In a dot density map satisfying Principle 1, if a more global layer is composed of polygons that contain the randomizing layer in a nested fashion, then there is no problem of overlapping regions and possible confusion about the assignment of dot to polygon. Thus,

Principle 2: Optimization principle
A nested hierarchy of layers provides optimized assignment of dots to more global layers.

Often, unfortunately, one is not able to obtain data arranged in a nested spatial hierarchy, such as the state and county units, tract and block group units, or another nested hierarchy. It may be desired, for example, to use Census data to obtain information about zip code polygons, school districts, minor civil divisions, and so forth. Census boundaries, however, are not generally commensurate with zip code boundaries and various other spatial units. In cases such as this, numerous creative approaches may be needed to align data sets to make appropriate comparisons.

In creating a dot density map, randomizing at the most local level is often, but not always, best. The reason it may not be best often involves zooming out so much that the highly localized randomization introduces clutter into the map—the dots create one big blob on the map. Conceptually, to reduce dot clutter, one might randomize at a more global level or one might alter what the dots represent. Generally, it is best to alter what the dot represents, instead of its position.

Principle 3: Scale principle
When a change in scale produces dot clutter from dot density, randomize at the most local scale available and alter dot representation to retain a dot density map satisfying Principles 1 and 2.

Map projections on which a geometric unit square represents the same amount of geographic area, independent of position, are called "equal area projections." On an equal area projection, relative land mass sizes appear as they do on the globe: Brazil is larger than Greenland by nearly 21 times, but on maps that are not in equal area projections, Greenland appears larger, for example. Dot density maps can be used to make comparisons. Unit (or other) squares represent the same amount of area on the Earth and comparisons at different latitudes are valid.

Principle 4: Projection principle
A dot density map must be based on an equal area projection.

If any projection other than an equal area projection is employed in making a dot density map, then comparisons involving area will be in error. This principle is of particular importance in constructing global maps and is less important in constructing local maps.

Dot density maps offer one way to synthesize information derived from point sources to suggest information about areas. As we shall see in future chapters, there are others, as well.

In the next section, you will have the opportunity to put these concepts into practice. The first activity involves changing the map scale in a web-GIS from the US Census Bureau. The next activity involves the analysis of population change using Esri web-GIS tools. In both activities, you will see the effect that map scale and symbology have on the accurate interpretation of the data.

5.4 Practice using selected concepts from this chapter

5.4.1 Scale change exercise

Access the TIGERweb mapping service from the US Census Bureau (http://tigerweb.geo.census.gov/tigerweb/). Consider the sequence of maps shown in **Figures 5.5 through 5.7** and focus on the Lake Michigan coastline that lies in the state of Illinois as the eastern edge of the greater Chicago metropolitan area. At a national scale, **Figure 5.5**, the coastline appears to be very smooth. The map shows a representative fraction of 1:18:489, 297 and it displays a bar scale.

Using the TIGERweb service, zoom in to a level that focuses on the southern end of Lake Michigan, adjacent to four states: Wisconsin, Illinois, Indiana, and Michigan (**Figure 5.6**). The representative fraction of this map is 1:1,155,581. This fraction is larger than the fraction associated with **Figure 5.5** so the map scale of the map in **Figure 5.6** is larger than that in **Figure 5.5**. Also note that county boundaries have now entered the picture as a consequence of

Figure 5.5 *Representative fraction: 1:18,489,297. Source: US Census Bureau, TIGERweb (beta), http://tigerweb.geo.census.gov/tigerweb/*

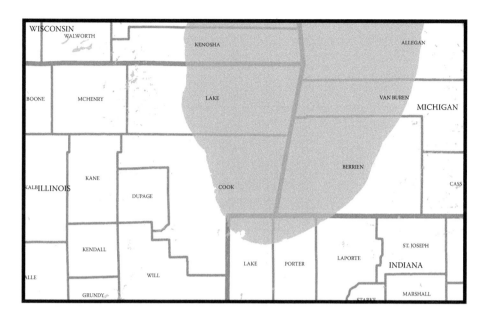

Figure 5.6 *Representative fraction, 1:1,155,581. Source: US Census Bureau, TIGER web (beta), http://tigerweb.geo.census.gov/tigerweb/*

zooming in. TIGERweb (US Census Bureau) has visibility of layers that is scale dependent. There is an assigned data range within which a given attribute is visible. In regard to the coastline of Lake Michigan, a bit more detail has come into focus but still it is quite smooth.

Zoom in even more (**Figure 5.7** with a representative fraction of 1:288,895). Again, extra detail enters the picture as the scale becomes more local than in **Figure 5.6**. Note the structures jutting out into the lake: Promontories and piers, bays and inlets—natural and man-made structures (can you tell which is which?). The man-made structures should be more angular. Remember that much of Chicago's Lake Michigan shoreline has been altered; even much of the eastern portion of the downtown is built on land filled and reclaimed from the lake (Mayer and Wade, 1973)!

If you were measuring the coastline length, and were using string, do you see that you would need much more string when the view is detailed than when it is general? Measurements, too, are often scale-dependent (Mandelbrot, 1983). As the map becomes more detailed, the length of the coastline increases, because more irregularities of the coast become visible. Thus, even the coastline of a relatively smooth-shored lake such as Lake Michigan, at an extremely detailed scale, becomes an almost impossibly large number to measure.

After trying this exercise using TIGERweb (**Figure 5.7** gives QR code access), try one of the Esri-based exercises, or another exercise that is linked at the

Figure 5.7 *Representative fraction: 1:288,895. QR code access to TIGERweb (beta), US Census Bureau. Source: US Census Bureau, TIGERweb (beta), http://tigerweb.geo. census.gov/tigerweb/*

end of the chapter. Now, imagine displaying the entire national map in **Figure 5.5** at the level of detail of **Figure 5.7**. Would the resulting map be large? Would it be easy to fold? Its large size would surely be impossible to fold. Can you think of reasons, now, why mapmakers have hard decisions to make in terms of deciding how much detail should enter the picture? As you can visualize, making decisions about the content of any map is complex. While GIS technology makes it easier to add and delete content, the decision as to what content to include is no less difficult today than in the past. One might argue that it is even more difficult because we have so many more types of data at our fingertips to map. What are the implications for accuracy and precision in measurement on maps at different scales? The implications for accuracy and precision in measurement are also just as important now as in the days of paper mapping. In fact, the danger of mis-interpretation is enhanced by the modern ability to zoom in to very large scales, even with data that were created at smaller scales. The features at that detailed, large scale appear to be located accurately, but they may have been plotted at a much smaller scale, and thus may be better symbolized with an aura of uncertainty as some cartographers have proposed (Clarke and Teague 1998).

5.4.2 Dot density maps: Investigating population change

Examine the map of future population growth and decline (http://esriurl. com/popchangedotmap) and view the legend while looking at Denver, Colorado. Note that the dot color depends on whether the area is expected to experience high growth, normal growth, or population decline through 2015. Do you find this map easy or difficult to interpret? Why? The numerous dots may make it difficult to interpret, but the dots also show trends that may be hidden by a standard choropleth map. How many persons does each dot, at this scale, represent? At this scale, one dot represents 180 people. Change the scale. How does the number of persons per dot change at different scales, and why? As you zoom in to larger scales, each dot represents fewer people, because you are now examining a smaller area. Click any area on the map and observe the growth or decline for the area you have selected from 1990 through 2015 as a bar chart. What patterns do you notice, and what are some reasons for those patterns? What side of Denver (north, east, south, or west) is projected to grow the most rapidly through 2015? It appears that the south and eastern sides of Denver are projected to grow most rapidly. The west side of Denver's growth is hindered by steep mountain terrain. Pan the map to Detroit. What do you notice about population change in Detroit versus in Denver? Much of Detroit is losing population, unlike Denver.

This tool also illustrates fundamental concepts of interpreting dot density maps that were discussed in this chapter. Reload the dot density population growth and decline map, which resets to Denver (**Figure 5.8**). Zoom in to the text on the map identifying the city of Lakewood, just west of Denver. Zoom in until you find some lakes just south of Lakewood; for example, Marston Lake, shown in the following figure. If you zoom in to a certain scale, you will see dots in the lake. Do these dots represent people living on houseboats? Why are dots shown in the lake? The dots are there not because people are living in or on the lake, but because the dots are randomly assigned to the statistical area that was used, such as census tracts or block groups. The dots do not represent the actual location where people live.

Examine the essay "Better Census Maps with Dot Density" on http://thewhyaxis. info/census/. The author reinforces some of the points we are raising in this chapter about the care that needs to be taken in generating any map, including dot density maps. Which of the dot density maps by race and ethnicity do you find most visually appealing? Why? How might these maps be misleading in terms of the amount of segregation by neighborhood in the cities depicted? Great care must be used in interpreting any map. What data were used to color code the dots? The maps give the appearance of more segregation than what might actually exist. Can you suggest ways to improve these maps?

Figure 5.8 *Dot Density Maps: Reflections on Best Practices. Source of base map: Esri software. Designer: Jim Herries, Esri.*

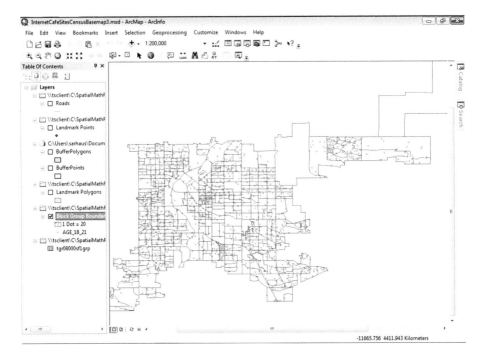

Figure 5.9 *Dot density map of 18 to 21 year-olds with dot scatter randomized within Block Groups. Block Group boundaries are shown. Source of base map: Esri software.*

5.4.3 Creating your own dot density maps: Exercise

Try to work the part of the Denver Internet Café Location exercise having to do with dot density mapping on your own. Use the .mxd file created in working with material at the end of **Chapter 3**. Join the downloaded Census .dbf to the Block Group shape file. The idea of a "join" in the world of GIS is parallel to the idea of "union" in the world of set theory. When two tables are joined, the new table contains all the entries of both; hence, the set of entries in the new table is the union of the set of entries in each of the two original tables.

Then, double-click the block group layer to open it and navigate to Symbology. Select the dot density option under "Quantities" and create a map for the 18 to 21 year-old category. Also create a choropleth map (choose "Graduated color"). **Figure 5.9** shows such a dot density map with the dot scatter randomized at the Block Group level. **Figure 5.10** shows the same map with the Block Group boundaries removed so that the clusters of dots are evident at the city-wide level. If you tried to construct this map on your own, where did you have difficulty? How does the dot density map of the 18 to 21 year-olds compare to the choropleth map (**Figure 5.11**) in terms of conveying meaning?

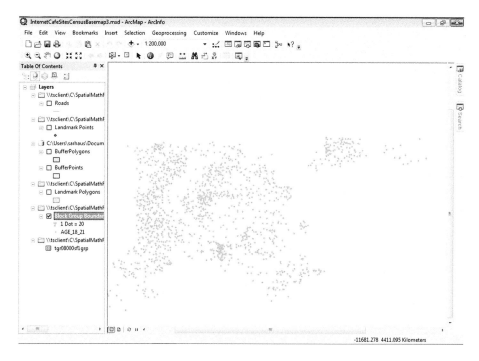

Figure 5.10 *Dot density map of 18 to 21 year-olds with dot scatter randomized within Block Groups. Block Group boundaries are removed to foster cluster visualization at the city-wide scale. Source of base map: Esri software.*

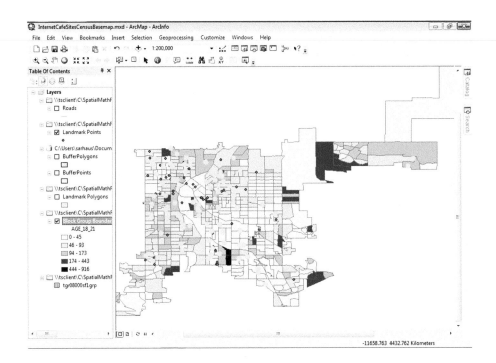

Figure 5.11 *Choropleth (graduated color) map of 18 to 21 year-olds, shaded by Block Group. Source of base map: Esri software.*

Experiment with the size and color of the dots. Next, vary the number of dots per person aged 18 to 21 years. Which size, color, and number of dots convey the meaning of the map most clearly? How can you best communicate the message to your map readers that the placement of the dots does not indicate where the 18 to 21 year-olds actually live within each block group? As you can see, with a desktop GIS such as ArcGIS for Desktop, you have much more flexibility in changing the symbology than with web-based GIS. Think about what other information you might wish to have. Then, go on to the future chapters where this problem will be revisited in greater detail. Often, it is a good idea to read a problem in advance and think about it broadly and then to revisit it once more information is known, bearing in mind that what was learned earlier (as in this chapter and in **Chapter 3**) might well be important in future activities.

5.5 Related theory and practice: Access through QR codes

Theory

Persistent archive:

University of Michigan Library Deep Blue: http://deepblue.lib.umich.edu/handle/2027.42/58219

From Institute of Mathematical Geography site: http://www.imagenet.org/

Arlinghaus, S. L. 2005. Spatial Synthesis. The Evidence of Cartographic Example: Hierarchy and Centrality. *Solstice: An Electronic Journal of Geography and Mathematics.* Volume XVI, No. 1. http://www-personal.umich.edu/~copyrght/image/solstice/sum05/dotdensitymap/overviewl.html

Arlinghaus, S. L. 1990. Scale and Dimension: Their Logical Harmony. *Solstice: An Electronic Journal of Geography and Mathematics.* Volume I, No. 2. http://www-personal.umich.edu/~copyrght/image/solstice/sols290.html

Practice

From Esri site: http://edcommunity.esri.com/arclessons/arclessons.cfm

Kerski, J. Studying Scale. Complexity level: 1, 2. http://edcommunity.esri.com/arclessons/lesson.cfm?id=490

Kerski, J. Siting an Internet Café in Denver Using GIS. Complexity level: 3. http://www.mylovedone.com/Kerski/Denver.pdf

Kerski, J. Studying Scale. Complexity level: 3a. http://edcommunity.esri.com/arclessons/lesson.cfm?id=496

Kerski, J. Studying Scale. Complexity level: 3b. http://edcommunity.esri.com/arclessons/lesson.cfm?id=502

Kerski, J. Studying Scale, ArcGIS version. Complexity level: 3. http://edcommunity.esri.com/arclessons/lesson.cfm?id=522

Kerski, J. Analyzing Demographic Components and Implications. Complexity level: 2, 3, 4, 5. http://edcommunity.esri.com/arclessons/lesson.cfm?id=714

6

Partitioning of Data
Classification and Analysis

Keywords: Data (62), map (43), method (27), population (22), classification (19)

> Unbiased at least he was when he arrived on his mission,
> Having never set eyes on the land he was called to partition
> Between two peoples fanatically at odds,
> With their different diets and incompatible gods.
>
> **W. H. Auden**

6.1 Introduction

Maps show locations; in addition, they often show themes. Themes can be ways to conceptualize the Earth, composed of biomes or watersheds, for example, or a series of events, such as a theme of the Napoleonic wars, or a specific variable, such as median age by country or by census tract, or in myriad other ways (Tufte, 1990). **Figure 6.1** shows a "thematic" map. In it, a set of data, or theme, has been partitioned into intervals. In theory, there are an infinite number of ways available to partition each theme. In practice, a finite set is typically employed. The method one chooses to partition the data, into mutually exclusive ranges, has a profound impact on how the resulting map appears and how it is interpreted (Monmonier, 1993, 1996).

6.2 The choice of data ranges

Different schemes of partitioning data can lead to vastly different maps. The maps in **Figures 6.1 through 6.5** are all maps of the world. There are four classes, or data ranges, in each map. The variable (attribute) being mapped is "area"—that is, land area. The purpose of taking care in selecting a ranging method is twofold:

- To facilitate the reading and understanding of the map.
- To reveal information that is not otherwise self-evident.

Consider the sequence of maps (**Figures 6.1 through 6.5**) that shows land area by nation partitioned into four classes (ranges), which are then mapped using a scale of reds. In this sequence, notice the differences in the appearance

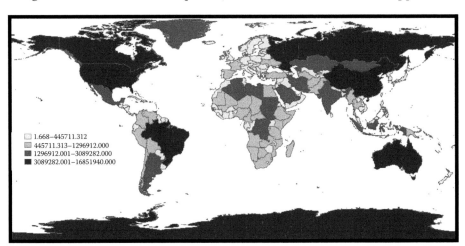

Figure 6.1 Land Area (square kilometers) partitioned by Natural Breaks. Source of base map: Esri software.

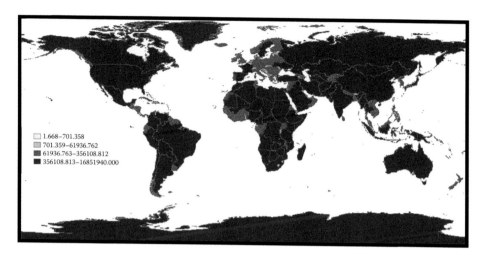

Figure 6.2 *Land Area (square kilometers) partitioned by Quantile. Source of base map: Esri software.*

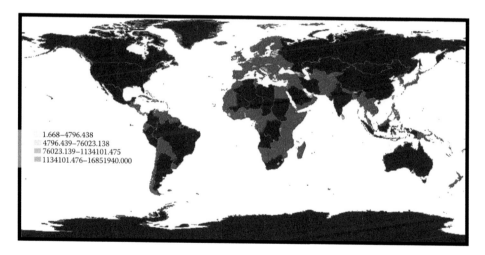

Figure 6.3 *Land Area (square kilometers) partitioned by Geometrical Interval. Source of base map: Esri software.*

of the maps. They all use the same underlying database. The reason they look different from each other is that the methods of partitioning the data into classes differ from map to map. Clearly, the method chosen for the partitioning of data is critical.

Think back to the chapter on color. In this map set, we use varying shades of a single color, rather than four different colors, so that the intensity of color reveals the intensity or the size of the underlying data. Once one has worked at deciding how to partition the data, then he or she also needs to consider how factors such as color selection might feed into such decisions.

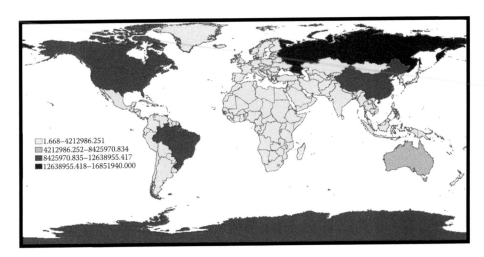

Figure 6.4 *Land Area (square kilometers) partitioned by Equal Interval. Source of base map: Esri software.*

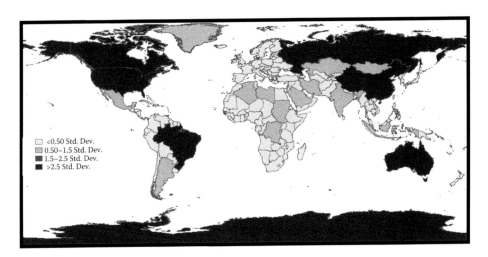

Figure 6.5 *Land Area (square kilometers) partitioned by Standard Deviations. Source of base map: Esri software.*

How is one to choose one method for partitioning data over another? After we discuss how each classification method works, we offer a simple catalogue of the merits and drawbacks of each (**Table 6.1**). In that discussion we indicate how to make such decisions for the methods of data partition presented here (there are others). However, the reader needs to be aware that there is no single recipe for making such selections; knowledge as well as common sense must both come into play. Readers with access to Esri software, such as ArcGIS, will find much useful material on classification available in the "Help" section of the supporting materials.

Table 6.1 Classification of Data

Method	Merits	Limitations
Natural breaks	Big jumps in the data appear at class boundaries. Extreme values are visually obvious. The two merits taken together may produce a "realistic" view of the data—hence, the possible suitability for choosing this method of data partition as a default.	Class intervals are difficult to read. Clear replication of results may be difficult. Merging or mosaicking maps will produce different classifications. Coloring may need to be adjusted so as not to give undue visual importance to extreme values.
Quantile	Well-suited to data that are linearly distributed—data that do not have disproportionate numbers of features with similar values ("clusters"). Easy to explain to others how it works. Useful for making comparisons in relation to the partition: For example, to show that a commercial establishment is in the top quarter of sales of all stores in the region. Distinctions among intermediate values, grouped in natural breaks, may be easier in quantile.	Features close in value to each other may lie in different classes. Features ranging widely in value may be included within the same class. Increasing the number of classes may help to overcome these drawbacks but that act then adds clutter to the map. Clear replication of results is easier than with natural breaks but still problematic when merging files.
Geometrical interval	Polygons that are largest in area are in classes by themselves.	Polygons that are smallest in area are grouped in classes and distinctions among them may be difficult to make.
Equal interval	Familiarity: A natural legend in terms of ease in reading (at least when the nature of the entire range of possibilities is clear, as in percentages, temperature, and so forth). Emphasis on ranking in relation to the partition: to show that a store is part of a group of stores in the top quarter in sales.	Hides variation between features with fairly similar values. When the data range does not already make natural sense, a different classification scheme may be better.
Standard deviation	Easy to visually assess which regions have values above or below the mean for all data.	The data may skew class count and position in relation to the mean: Many high values may cause low values to be grouped in a single class below the mean and produce multiple classes above the mean, so that the mean class does not, itself, occupy a visual central position on the map.

Generally, it helps to graph the data set in advance. Determine if it is linearly distributed. Determine if it has clusters near the top or the bottom of the distribution. Different partitioning methods work well in some styles of distribution but not in others. Choose a partition that is most appropriate for your data (Paret, 2012).

6.2.1 Natural breaks

This method is the default partitioning scheme in many contemporary mapping packages. It identifies breakpoints between ranges using a statistical formula (according to Jenks' optimization; de Smith, Goodchild, and Longley, 2012). Simply stated, this method minimizes the sum of the variance within each of the ranges. It attempts to find groupings and patterns inherent in the data set. Some might view it, for ease, as equivalent to throwing a pile of papers down the stairs and grouping the papers by step. The map in **Figure 6.1** shows the worlds' lands grouped by national area size into four different ranges with the partition created using Jenks' method.

6.2.2 Quantile

In the quantile data range partitioning method, each range contains the same number of observations. Note that the word "quantile" does not prescribe how many ranges are to be chosen. If four ranges are chosen in the quantile partition, those quantiles are known as "quartiles." If five ranges had been chosen, they would have been "quintiles." The word "quantile" is a derivative of the word "quantity." Quantiles are easy to understand, but they can be misleading. Population counts, for example, are usually not suitable for quantile partition because only a few places are highly populated. This distortion can be overcome by increasing the number of data ranges, but it is better to choose a more appropriate ranging method. Quantiles work well with linearly distributed data: Data sets that do not have disproportionate numbers of features with similar values. **Figure 6.2** shows such a partition for the sample data set here, that of land area by country. Most of the countries appear to be in the highest class (the class with the most land area), but this is simply because, at this scale, the largest countries are the most visible. The data are not linearly distributed: A few countries have large land areas, but most countries are relatively small. The same number of countries exists in each class, but only upon zooming in to a larger scale can most of the countries in the other three data classes be seen.

6.2.3 Geometrical interval

This method partitions data by finding breakpoints based on class intervals that have a geometrical series. The complex algorithm attempts to ensure that classes have similar numbers of values and that change between the classes is close to uniform. Ranges determined with the geometrical interval method are often very similar to quantile ranges when the sizes of all the features are roughly the same. Geometrical interval ranges will differ from quantile ranges if the features are of vastly different areas, as is the case in this data set (**Figures 6.2 and 6.3**). In earlier times, this method was known as "smart quantiles." This method works farily well on data that are not distributed normally. It is a method that is difficult to communicate via a legend or other summary materials.

6.2.4 Equal interval

The equal interval classification method partitions the range of attribute values into equal-sized subranges. Then the values are classified based on those subranges (**Figure 6.4**).

6.2.5 Standard deviations

When the data set is ranged using the standard deviations method, the software finds the mean value and then places range breaks above and below the mean at intervals of (usually) either 1/4, 1/3, 1/2, or 1 standard deviation until all the data values are contained within these ranges. The software should aggregate any values that are beyond three standard deviations from the mean into two classes, greater than three standard deviations above the mean (">3 Std Dev.") and less than three standard deviations below the mean ("<–3 Std. Dev."). **Figure 6.5** shows the world's land areas grouped according to this ranging method. In an effort to reinforce interactive thinking between maps and math, readers familiar with calculus might reasonably ask what, if any, role the Mean Value Theorem or the Intermediate Value Theorem might play in the classification method described here.

Merits and limitations of the different ranging methods to partition raw data are summarized (**Table 6.1**). What is critical is to understand that the choice of partition can produce vastly different analysis and interpretation. Thus, it is important in research to be clear why one ranging method was chosen over another. Indeed, it may often be prudent to display data mapped using several different ranging methods.

Which of the maps in **Figures 6.1 through 6.5** is "best"? In looking at the merits and limitations of each of this set of commonly used ranging methods, we see that the pattern of the underlying data may help us to answer this question. The answers that come from the data should make good intuitive sense when looking at the maps, themselves, based on our general knowledge. **Figure 6.6** shows the underlying attribute table of the map opened in Microsoft Excel and graphed as a simple line chart in Excel. (Or, one can employ the onboard classification histogram embedded in ArcGIS software as is done in one of the activities near the end of this chapter.) Clearly, the red line of the chart is *not* linear. Thus, one would expect that the quantile method of ranging of data would not be appropriate. That idea fits with the appearance of the map in **Figure 6.2**. That map conveys very little information. Thus, we would reject quantiles as a mapping classification method for this set of data.

Similarly, the Geometrical Interval method, while it makes more distinction among data entries than does the quantile method, it still does not show clear distinctions among the four classes. The entries appear bunched into two of the four classes. Partition by Equal Interval makes reasonable distinctions for the larger countries, but mid-sized and small countries appear grouped together in a single class. Partition by Standard Deviation is skewed, as is Equal Interval, to have more distinctions made among larger countries; however, the degree of skew is not as severe in the Standard Deviation partition as in the Equal Interval partition. Thus, of this set of maps, we would likely choose either Natural Breaks or Standard Deviation, depending on the application, as a method for partitioning this data set.

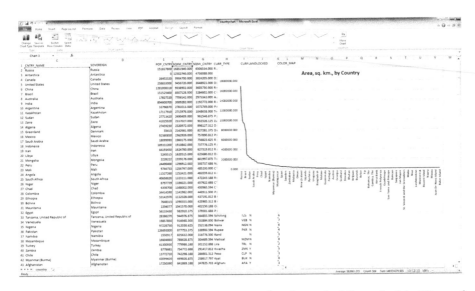

Figure 6.6 *Attribute table of maps in* **Figures 6.1 through 6.5** *graphed in Microsoft Excel.*

What is critical to understand here is the importance of considering how one classifies data sets in advance of presenting a final map. When the final map is presented, it is also necessary for the mapmaker to explain decisions that were made in regard to partitioning of data because the results can vary greatly depending on these fundamental decisions. Equally, it is critical for the map reader to ask questions about how data sets were partitioned in advance of accepting apparent spatial pattern presented on maps derived from underlying data sets—it is far too easy to gerrymander results—we should take to heart Monmonier's (1996) cautions in interpreting maps!

6.3 Normalizing data

In the previous section, we illustrated the importance of choosing an appropriate ranging technique to display data—bad choices lead to bad results. Of equal, but different, importance is the issue of normalizing data. The maps of the previous section were all based on total land area by country. Often, mapping totals might be an informative way to discover patterns in the data, in this case, mapping how many square kilometers each country occupies. With other types of data, especially with population data, merely mapping total counts may be insufficient.

Consider **Figures 6.7 and 6.8**. In **Figure 6.7**, the population of the United States is mapped by county and displayed as total counts by county. The classification chosen was Quantile in five ranges. In the region of California, Nevada, Utah, and Arizona, many of the counties appear quite large in land area; and by virtue of having a large land area, many of these counties also have relatively large populations. Perhaps that is not surprising; however, it does give a skewed picture of concentrations of population.

In **Figure 6.8**, we divide the total count by the area to measure population density, which might have more meaning in terms of seeing where there is concentrated population. Compare northern Arizona and Nevada in these two maps. Which one better fits what you already know about population distributions and concentrations in this part of the United States? The data in **Figure 6.8** have been "normalized." Normalizing data divides each of the attribute values by some other value. Use the QR code of **Figure 6.9** to link

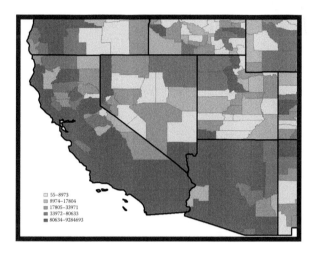

Figure 6.7 *United States population mapped by county, total count. Source of base map: Esri software.*

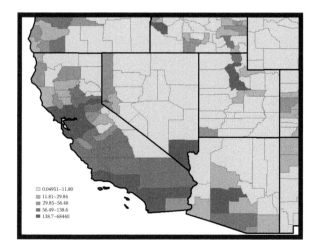

Figure 6.8 *Total population divided by area, per county. Normalized data. Source of base map: Esri software.*

Figure 6.9 *Video on normalizing data. Link to it via the QR code or the link below.*
http://www.youtube.com/watch?v=t5cG2D4Ln0E

to a video that will offer multiple examples, clearly explained, of how one might be creative with normalization. This topic is one that is deep and rich in content. The reader will benefit greatly from a live video with hands-on explanations. As with any mapping capability, there are merits and limitations to the normalizing of data (**Table 6.2**).

6.4 Inside, outside, wrong side around

A hidden, yet critical, theorem is necessary in mapping software if one is to be able to use software to accurately color polygons and assign addresses to streets. To do so, it is necessary for the mapping software to be able to distinguish the inside from the outside of a polygon.

Table 6.2 Normalization of Data

Merits	Limitations
Divide by the sum total of the attribute's values so that the resulting ratios represent percentages of the total. Enables comparison from one region to the next, using percentages of the total (region 1 contains 50% of the sales while region 2 contains a mere 17% of the sales) rather than absolute totals. Divide by values in another attribute: This procedure may take into account spatial variation influencing the original attribute. Population density, dividing total population per unit by area of the unit, is a common example.	If total count is important, then normalization of data is not appropriate. For example, it may be more important to know how many members of a minority group are present in a particular region, to trigger some funding mandate, than it is to know what the density of population is within that particular group. In a group of 100, 35 members of a minority group may appear fairly "dense"; however, if 50 members are required as a floor for certain programs to be realized, then the density is irrelevant. Do not normalize data that has already been normalized, such as rates, attributes per unit area, and so forth. A GIS makes it so easy to normalize data that caution must be used when determining what the denominator is to normalize data by. For example, in working with Census data, dividing the number of Asian Americans by the number of housing units does not yield the number of Asian American households. It is critical to understand each data element so that the results of the normalization are accurate. One can certainly divide "apples by oranges" but the results will not necessarily mean anything.

A simple closed curve is one that is topologically equivalent to a circle: Like a rubber band, it can be snapped back into a circular form, no matter how many bends it might have along its journey from start back to the same point as endpoint. The following theorem, from topology, about simple closed curves is attributed to Camille Jordan (1909).

> Jordan Curve Theorem: A simple closed curve J, in the plane, separates the plane into two distinct domains, each with boundary J.

Simple closed curves are often referred to as Jordan curves. The two separate domains are often called "inside" and "outside" to fit with intuitive ideas. To determine which the "inside" is, imagine walking along the curve in a counterclockwise direction. Your left hand points to the inside of the curve; you are functioning as a continuously turning line tangent to the curve. This view offers precision and replication of results to the otherwise intuitive notion of "inside." An example of a curve that is not a Jordan curve is a figure eight. Walk along it, counterclockwise, with your left hand pointing to the "inside." When you walk past the point where the curve crosses itself, what should be the "inside" has now become the "outside" (**Figure 6.10**).

To illustrate this idea in a real-world context, consider a portion of the Chicago street map in the Hyde Park (University of Chicago) neighborhood. We assume (contrary to fact but for the sake of demonstration) that all streets in this area are two-way streets. Suppose that you take a path from the dot placed at 57th Street and Woodlawn Avenue and travel north on Woodlawn to 56th Street, then west on 56th to University Avenue, then south on University Avenue to 57th Street, then east on 57th to Kimbark, then south on Kimbark to 58th Street, then west on 58th to Woodlawn, and then north on Woodlawn to the

Figure 6.10 *A curve that crosses itself is an example of a curve that is not a Jordan curve.*

starting point. This path is traced out in **Figure 6.11a**. Suppose that you have been directed to color the area to your left as a light red. The result appears in **Figure 6.11a** and it is probably not what was intended—instead the intention was to color what intuitively appeared to be the interior. The solution is simple. Break this single curve that crosses itself into two separate simple closed curves at the starting node. Then color the inside of each rectangle individually.

Another sort of unintended consequence that can arise from having polygons whose edges cross appears in **Figure 6.11b**. Now, the path is the same here, for the first four edges as in **Figure 6.11a**. But this time, when the starting node is returned to, turn south on Woodlawn and go around the block the other way. This path is indicated by the arrows in **Figure 6.11b**. In this area of Chicago, the even-numbered addresses are on the west side of the north–south streets and the

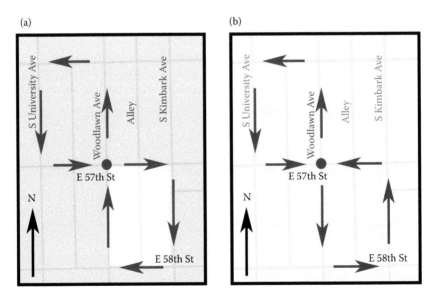

Figure 6.11 *(a and b) Unintended consequences coming from Jordan Curve Theorem failure. The red arrows trace out a curve similar to that in **Figure 6.10**.*

odd-numbered addresses are on the east side of the north–south streets. Suppose you have been directed to find an even address on Woodlawn, somewhere "near" 57th Street, on the left side of Woodlawn. If you begin at the node, head north on Woodlawn, look on the left side of Woodlawn and find it is not there (even though you are looking at even numbers that are perhaps too low), then proceed back past the node along the indicated path, as you look on the left, you will now be looking at the odd-numbered addresses. You will not be able to execute with accuracy. Again, the problem is solved by removing the node.

Thus, when working with digital maps, it is important to know that the digitizing process did not introduce any polygons that cross themselves, and, that any nodes that produce such crossings were removed in the digitizing process. The examples above are straightforward; it is not always that straightforward when actually digitizing a complicated street map. Fortunately, digital maps made since 1990 and the release of the first TIGER/Line files providing nationwide street centerline coverage of the United States (US Census Bureau, 2012), are free from this difficulty.

However, the USGS Digital Line Graphs (DLG), another commonly used vector data set, have no standard about the order of digitizing. Therefore, adjacent line segments may start and end in different orders and thus contain different directions as part of their attributes. When one wishes to select a certain polygon, therefore, through the use of these line segments, one needs to use statements like "select line segment A where the left side polygon = 1 OR the right side polygon = 1," to account for the different directions in which a line segment was digitized. Not including the "OR" statement can lead to a smaller number of polygons being selected, potentially skewing the resulting analysis.

Therefore, it remains relevant to understand the sort of unintended consequence that a lack of attention to underlying geometry, and order within that geometry, might cause. One must have the Jordan Curve Theorem built into the software if geocoding is to work on address, color, or other assignment problems matching data to spatial units. This process that involves "order" is critical, but unseen by most readers of maps.

Try your hand at the Esri geocoding exercise linked at the end of this chapter. Geocoding, in one form, is the process of associating geographic location to street addresses. It can involve, therefore, concepts such as "all of the even numbers lie along the west side of the street." As you work the exercise, think about what sorts of problems might have arisen if considerations involving digitizing order and the Jordan Curve Theorem had not been integrated into the software background.

6.5 Making something from nothing?

The Thiessen polygons in **Chapter 3** illustrated one way to create polygons from a scatter of points. There are a variety of others, as well, that permit the definition of regions in the absence of regional information. All rely on the concept of partition. We look here at a small selection of possibilities.

6.5.1 Isolines; contours

Thiessen polygons offer one way of defining regions in the absence of regional information. Still another way to partition the plane involves isolines. An isoline is a curve or a line on a map along which there is a constant value. It partitions a set of values as all the points on one side of the line are less than the constant value and all the points on the other side are greater than the constant value. When that value is based on topographic elevation, the isoline is called a contour. Isolines appear commonly on television weather stations representing temperature (isotherms), precipitation (isohyets), barometric pressure (isobars), and other weather phenomena.

From the standpoint of the mathematical connection, students of calculus will recognize isolines as so-called level curves—as in a set of cross sections at different levels of a surface. Slicing a cone with a circular base and apex directly over the center of the circular base yields circular cross sections of different sizes at different levels (heights above the base). When these are projected back into the plane, they form a set of concentric circles, similar in appearance to contours. Real-world contours representing a mountain on a map could be created by finding level curves of the mountain. To do so, however, would require knowing the equation of the surface/volume of the mountain. Mountains are not cones or any other straightforward three-dimensional mathematical structure. What is mapped, however, is a two-dimensional representation of a two-dimensional surface—the third dimension present in the real world is lost in the transformation. The purely mathematical approach, which affords accuracy, is difficult.

A lunar data viewer offers an interesting way to look at cross-sections. It can be accessed at http://target.lroc.asu.edu/da/qmap.html. Pick a crater. Draw a line across it and a nice cross-sectional graph will appear. How wide and deep are some of these craters? The largest ones are just over 500 kilometers wide and 4000 meters deep. Use caution when interpreting the cross sections from this or any website, paying attention to the horizontal and vertical units of measurement. Speculate on angle of impact, relative age of the crater, and size of the meteor! Spatial thinking lends one to the conclusion that the newer impacts are those that are superimposed on other, older, landforms. The lunar maria, formed from basalt flows from volcanoes, have in many locations obliterated the former surface, so despite their old age, they too are younger than the landscape underneath.

In simple plane geometry, there are other strategies to approximate contour location that are easier to implement than finding level curves and that rely only on simple partitioning of data. The example below suggests the general strategy.

(a)   (b)

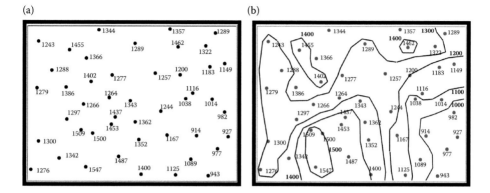

Figure 6.12 *(a) Scatter of weighted dots. (b) Partition of dot scatter into mutually exclusive sets (done by hand for illustrative purposes). (Arlinghaus, S. L., unpublished lecture notes.)*

Consider a scatter of dots with weights attached (often elevation); **Figure 6.12a**. Use a line to partition the scatter into three mutually exclusive and exhaustive sets (**Figure 6.12b**):

- All values on the line are identical.
- All values on one side of the line are less than the value along the line.
- All values on the other side of the line are greater than the value along the line.

There are an infinite number of ways that the contours can be placed between nearest neighbors—via a simple linear split or perhaps via some sort of split representing concavity of the surface, or distance decay. There is also the "spline" or "rubber sheeting" method that seeks to minimize total curvature. Also, contours can be run off the edge of the page instead of wrapping across, as does the 1300 contour in **Figure 6.12b**. The process of approximating contours in the plane is not difficult, using this sort of simple partitioning scheme, but it is complex. **Figure 6.13** shows a professionally drawn topographic map with a complicated pattern of contours; the juxtaposition of the hand-drawn with the professionally crafted map illustrates how a simple partitioning procedure in the plane becomes complex.

Try some of the accompanying exercises at the end of the chapter! Learn about the importance of slope, not only in skiing but also in mapping. Study a variety of the linked exercises to examine important real-world concepts related to contours and slope such as steepness, hill shading, aspect, and orientation.

6.5.2 Mapplets

Maps portrayed as "applets"—hence, "Mapplets" may offer a visual display for identifying breakpoints in data, although one must exercise thoughtful

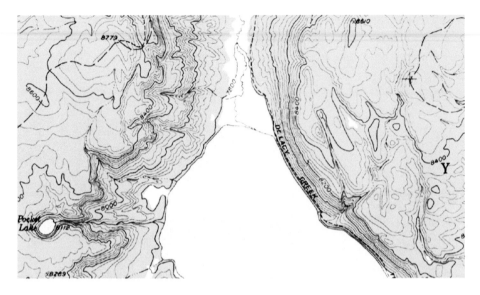

Figure 6.13 This image shows a sample of a professionally made map with contours. Closely spaced contours reflect steep slopes. Source: USGS, The National Map, US Topo, http://nationalmap.gov/ustopo/index.html

care in interpretation of visual evidence reconstructed from numerical pattern alone, without regard to the real-world context (Monmonier, 1996). **Figure 6.14** shows a structural model of Varroa mite (pest attacking honey bees) diffusion through time (Sammataro and Arlinghaus, 2010). In the online version (Java-enabled browser), pull the red year box for 1963 all the way across to the right (use the scroll bar) and then drag and drop various pieces of the left side of the mapplet to unravel it and see the pattern of possible time points of intervention opportunity, data breakpoints, at various stages in the diffusion process. Where clipping out a single box would disrupt the flow, that location might suggest intervention opportunity. If a Mapplet box "sticks" on another, pull it a bit in a different direction. Generally it is possible to move beyond the obstacle. Mapplets seem to offer a wide array of possibilities for description, interpretation, and analysis of complex spatial systems. Click on the link, in the References below, to see more of the Mapplet than can be displayed in the static image (Arlinghaus, 2000).

The idea of "partition" is one that is rich and varied. Readers interested in more conceptual material on this topic and on related topics might consider reading works on location theory (Haggett, 1977). Try von Thünen (1966 translation) and agricultural location; try Weber (1928) and industrial location; or, try a more current work such as the one by Puu (1997).

In the next section, you will have the opportunity to put selected concepts from this chapter into practice, and classification in particular, first with a web-based Geographic Information System (GIS) and then with a more powerful desktop-based GIS.

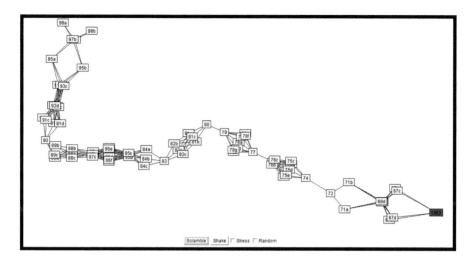

Figure 6.14 *Mapplet of Varroa mite diffusion. Source: Arlinghaus, S. L. (from input from D. Sammataro). 2000. Animaps IV: Of Time and Place, Solstice: An Electronic Journal of Geography and Mathematics, Volume XI, No. 1. Ann Arbor: Institute of Mathematical Geography. http://www-personal.umich.edu/~copyrght/image/solstice/ sum00/animapsiv.html.*

6.6 Practice using selected concepts from this chapter

6.6.1 Investigate classification using ArcGIS online

In this activity, you will use web-based GIS to investigate classification. The objectives of this activity are to: Understand and use different classification methods; understand how classification is related to data; make maps using different classification methods; understand how and why those methods differ, and use them in conjunction with classification (standard notation, mean, standard deviation).

Start a web browser and access ArcGIS Online (www.arcgis.com/home). In the search box, search for "Oklahoma 1900 to 2010 Population Change owner:jjkerski," and open it with the ArcGIS Online map viewer. If while opening the map, you are prompted to use ArcGIS Explorer Online, decline this choice; that is, open it in the ArcGIS.com map viewer. You should see Oklahoma counties in the same uniform green symbol (**Figure 6.15**). Click on the button allowing you to show the contents of the map, and click on the Oklahoma Counties 1900 to 2010 layer. Find the arrow pointing to the right, and use this arrow to access a context menu. From the content menu, select "Change Symbols." Change the contents to the right of the "Use" text to "Color" to make a color thematic map, and select the field "Pop 1930" to map the 1930 population by county. Start with natural breaks and apply your changes. What spatial patterns of population do you notice around the

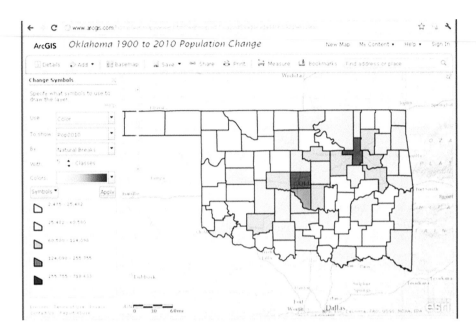

Figure 6.15 Oklahoma population change over time. Source of base map: Esri software.

state in 1930? While Oklahoma and Tulsa counties are prominent, the site of Oklahoma City and Tulsa, the largest urban centers in Oklahoma, many rural counties have fairly high population as well, right before the Dust Bowl caused a mass exodus. What is the total range of the data? The data ranges from 5408 to 221,738.

Change the method of thematic mapping to equal interval and note the pattern. Repeat the process for standard deviation, and then for quantile. Which classification method gives you the best understanding of the total population and the spatial distribution of the population for 1930? You are the data producer and the data consumer, and you have the power to make the decisions, but keep in mind the tenets introduced in this book, and in particular, the understanding your map imparts to the end user. Next, make a natural breaks map for 1940, and then for 1970, and then for 2010. How does the spatial pattern change over time? What patterns are persistent over time? Oklahoma City and Tulsa increase in population while rural areas decline, resulting in a less dispersed and more nodal structure. The rural counties that do increase in population consistently are suburbanizing counties near the population centers. What do you predict Oklahoma's population totals and patterns will be in 2050? The trends since 1940 are likely to continue; that is, rural population loss and continued urbanization. Note the numbers at the low and high end of the ranges for each of the above years 1930, 1940, 1970, and 2010. What does the fact that the population of the least populous county in the state decreases, rather than increases, tell you about changes in Oklahoma's population over the years? This indicates that sparsely populated rural counties, with agricultural mechanization and the loss of small farms, will continue to lose people. One county in the Oklahoma panhandle did post recent increases, but it is due to large feedlots, a facet of agribusiness.

6.6.2 Digging deeper into classification using ArcGIS for desktop

In this activity, you will again use GIS to investigate classification, but this time, dig deeper than you could with the online tools, this time with ArcGIS for Desktop. The software employed will be ArcGIS for Desktop, from Esri. Start ArcGIS for Desktop and access the ArcMap application. Use File→ Open to open the lesson_high4_classification .mxd file which may be obtained from the linked website, http://edcommunity.esri.com/arclessons/lesson. cfm?id=523

Remember that each method of classification changes the meaning conveyed by your map to the map reader, so you need to choose the classification with care. The steps in this activity will lead to the appearance of a map similar to that shown in **Figure 6.16**.

Which map layer is already classified? The elevation layer is already classified. How many classes of elevation are there and what is the range of the elevation data? What are the units of elevation data? The elevations are in meters. Label countries based on CNTRY_NAME. Use Identify on the elevation layer to answer the following: What is the Elevation code for the class corresponding to the highest elevation? The code is 16, grouped in the category 13 through 16. As shown in **Figure 6.17**, right or double click on the Elevation map

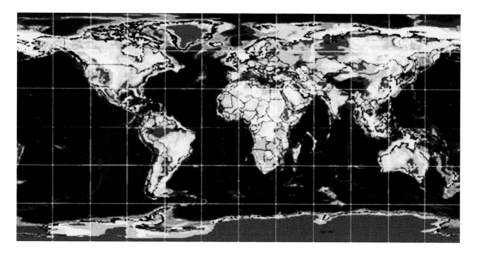

Figure 6.16 *Classifying data on maps using Esri software. Source of base map: Esri software.*

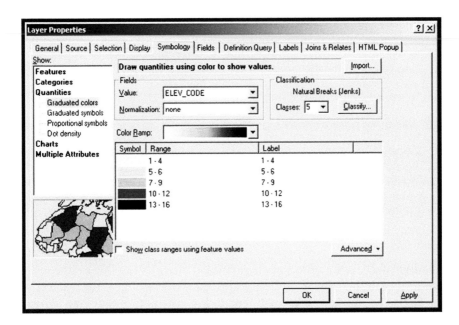

Figure 6.17 *Elevation Code layer. Classifying data into ranges using Esri software. Source of base map: Esri software. Look at elevation on other surfaces, such as a lunar map with high resolution patches: http://target.lroc.asu.edu/da/qmap.html.*

layer → Properties → Symbols → Change to Graduated Symbols → change the Value to Elev Code → OK.

Change the symbology to Unique Value and observe your map. Change the symbology back to Single Symbol. What is the difference between maps symbolized by: Single Symbol, Unique Symbols, and Graduated Symbols? The map symbolized as single symbol as well as unique symbols make it impossible to interpret the elevation values. A middle ground is represented by graduated symbols that show elevation as a series of about five categories.

Turn off Elevation. Change the symbols of the Country map layer to Graduated Colors on Field (value) POP2005 (Population in 2005). Countries will be classified under the Natural Breaks classification method. In the Symbology window, click on Classify and observe the histogram which will appear as in **Figure 6.18**. Why are there so many records in the first class? So many records exist because most countries in the world actually have a small population. What is the mean population of world countries? The mean population is 25.8 million. What is the Standard deviation of the population of world countries? This value is 111.8 million. What is the total world population according to this data set? This is 6.4 billion according to this data set, and because the world population now is 7.2 billion, this data set is obviously out of date.

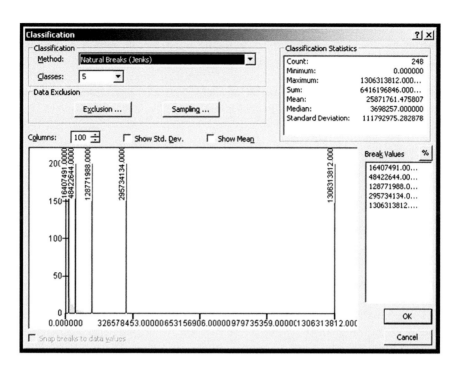

Figure 6.18 Histogram, associated with map, made using Esri software.

Expressed in "standard" rather than in "exponential" notation, note the pattern of distribution of the mean and standard deviation of the population of world countries—Mean: 2.58 E 7 = 25871761.475806452, or about 25.8 million people; Standard Deviation: 1.11 E 8 = 111792975.28287828, or about 111.7 million people. Most of the countries on this map will fall into the lowest category with the exception of China, India, and the USA. Maps such as this are not very helpful except for examining outliers.

Use "Classify" and change to Quantile, again noting the histogram. Is this quantile map more helpful than the Equal Interval map? Change the Symbols to Manual. What manual class breaks make the most useful map? How many classes do you believe are optimal? Again, you are the map producer and map reader, so you have the power to make the choices. But, most studies have shown that people can easily process between six and nine classes, after which it becomes much more difficult to distinguish the classes. Of Equal Interval, Quantile, or Manual, which classification method was most useful for this population data? Quantile, as explained in this chapter, is not as useful to use at this scale and with population data, because most of the small countries at a global scale will not be visible.

Quantile places the same number of observations into each category (by "Quantity"). Here, with 248 countries and five classes, about 49 or 50 countries fall into each category. In the above activity, the Quantile map did a better job of picking out pattern than did the Equal Interval map. On the other hand, larger countries in size tend to have more population, so at this scale it is difficult to see the pattern of the smaller-population countries. There is a balance between enough classes to differentiate between countries, but not too many or else it exceeds the mind's capacity to differentiate between the classes and colors. Thoughtful analysis in choosing data ranges is critical; vastly different visual displays may result from the same data set based on choice of ranging the data!

6.6.3 Normalization activity

Return to the Denver Internet Café activity that you began in an earlier chapter. Now, focus on the normalization section of the activity. Normalization allows you to see patterns that you cannot see while mapping total quantities, such as the percentage of 18 to 21 year-olds. Compare the raw values mapped in **Figure 6.19** (from **Chapter 5**) to the normalized values (with number of 18 to 21 year-olds divided by total population in 2000) for each Block Group (**Figures 6.20 and 6.21**). One way to see the number of block groups with more than 10% 18 to 21 year-olds is simply to alter the way in which the data set is partitioned, as in **Figure 6.22**. Try it! As in the previous chapter, consider trying more of the activity for yourself, to return to a more complete view of it in a later chapter.

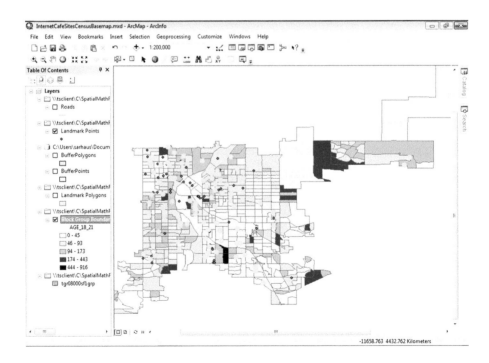

Figure 6.19 *Raw values of 18 to 21 year-olds mapped by Block Group. Source of base map: Esri software.*

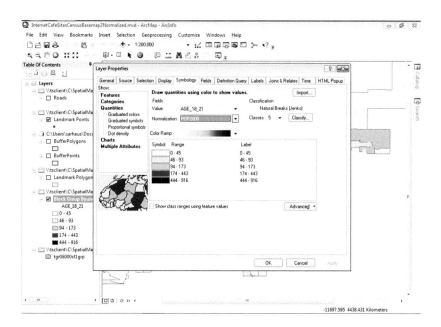

Figure 6.20 *Normalize data in the Symbology window. Here, a choice of Census data for 2000 will be totals by Block Group because those are the underlying spatial units. Source of base map: Esri software.*

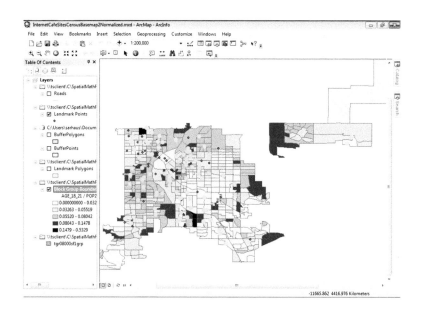

Figure 6.21 *Block Group data on numbers of 18 to 21 year-olds as a fraction of total Block Group population. Data are normalized by dividing total number of 18 to 21 year-olds by total population, by Block Group, for each Block Group. Source of base map: Esri software.*

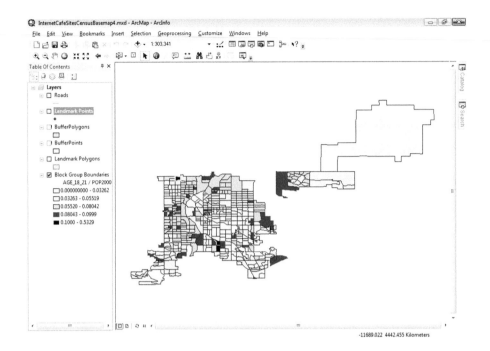

Figure 6.22 *Darkest block groups are those with 10% or more of its population in the age range of 18 to 21 years. Source of base map: Esri software.*

6.7 Related theory and practice: Access through QR codes

Theory

Persistent archive:

University of Michigan Library Deep Blue: http://deepblue.lib.umich.edu/handle/2027.42/58219

From Institute of Mathematical Geography site:http://www.imagenet.org/

Sammataro, D. and S. L. Arlinghaus. 2010. The Quest to Save Honey: Tracking Bee Pests Using Mobile Technology. *Solstice: An Electronic Journal of Geography and Mathematics.* Volume XXI, No. 2. http://www.mylovedone.com/image/solstice/win10/SammataroandArlinghaus.html

Arlinghaus, S. L. 2000. Animaps IV: Of Time and Place. Works Best with Microsoft Internet Explorer. *Solstice: An Electronic Journal of Geography and Mathematics.* Volume XI, No. 1. http://www-personal.umich.edu/~copyrght/image/solstice/sum00/animapsiv.html

Arlinghaus, S. L., W. D. Drake, J. D. Nystuen, A. Laug, K. S. Oswalt, D. Sammataro. 1998. Animaps. *Solstice: An Electronic Journal of Geography and Mathematics.* Volume IX, Number 1. http://www-personal.umich.edu/~copyrght/image/solstice/sum98/animaps.html

Practice

From Esri site: http://edcommunity.esri.com/arclessons/arclessons.cfm

Kerski, J. Investigating Classification. Complexity level: 1, 2.

Kerski, J. Siting an Internet Café in Denver Using GIS. Complexity level: 3. http://www.mylovedone.com/Kerski/Denver.pdf

Kerski, J. Investigating Classification. Complexity level: 3a. http://edcommunity.esri.com/arclessons/lesson.cfm?id=497

Kerski, J. Investigating Classification. Complexity level: 3b. http://edcommunity.esri.com/arclessons/lesson.cfm?id=503

Kerski, J. Investigating Classification. ArcGIS version. Complexity level: 3. http://edcommunity.esri.com/arclessons/lesson.cfm?id=523

Kerski, J. Extreme Temperatures in USA Investigation. Complexity level: 2, 3, 4. http://edcommunity.esri.com/arclessons/lesson.cfm?id=576

Kerski, J. Learning about Local Water Quality w/ArcGIS Online. Complexity level: 2, 3, 4. http://edcommunity.esri.com/arclessons/lesson.cfm?id=708

Kerski, J. GIS Lesson Demonstrating Local Maxima. Complexity level: 3. http://edcommunity.esri.com/arclessons/lesson.cfm?id=664

Kerski, J. Ranges and Extremes. Complexity level: 3. http://edcommunity.esri.com/arclessons/lesson.cfm?id=528

Kerski, J. The World War II Journey of the Pacific Clipper. Complexity level 3, 4. http://edcommunity.esri.com/arclessons/lesson.cfm?id=539

Kerski, J. Siting a Ski Area: Continental Divide. Complexity level: 3, 4, 5. http://edcommunity.esri.com/arclessons/lesson.cfm?id=622

Kerski, J. Siting a Ski Area: Direction of Slope. Complexity level: 3, 4, 5. http://edcommunity.esri.com/arclessons/lesson.cfm?id=629

Kerski, J. Siting a Ski Area:Slope and More, A. Complexity level: 3, 4, 5. http://edcommunity.esri.com/arclessons/lesson.cfm?id=630

Kerski, J. Siting a Ski Area: Slope and More, B. Complexity level: 3, 4, 5. http://edcommunity.esri.com/arclessons/lesson.cfm?id=631

Kerski, J. Siting a Ski Area: Webcams, Elevation. Complexity level: 3, 4, 5. http://edcommunity.esri.com/arclessons/lesson.cfm?id=632

Kerski, J. Siting a Ski Area: Synthesis. Complexity level: 3, 4, 5. http://edcommunity.esri.com/arclessons/lesson.cfm?id=633

Kerski, J. Exploring World Demography through GIS. Complexity level: 4. http://edcommunity.esri.com/arclessons/lesson.cfm?id=605

Kerski, J. Geocoding Using ArcGIS Online. Complexity level: 5. http://edcommunity.esri.com/arclessons/lesson.cfm?id=440

7

Visualizing Hierarchies

Keywords: Data (43), hexagons (36), map (26), block (25), hierarchy (24), area (22)

> Mapmaker, mapmaker, make me a map.
> Find me a find, catch me a catch
> Mapmaker, mapmaker, look through your book,
> And make me a perfect map.

Modified from *Fiddler on the Roof*

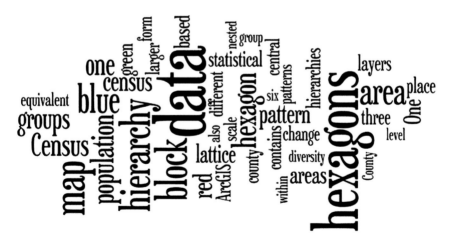

7.1 Introduction

The concept of hierarchy is important in the real world, in the virtual world, and in all worlds where the two interact. Generally, a hierarchy is an ordered set of items arranged in a fashion where it is possible to state, with certainty, that any given item is "above," "below," or "at the same level as" any other element in the set. Most military organizations have a clear-cut hierarchy of ranks of officers. Often churches, universities, and other institutions have hierarchies of administrators or professional staff. The certainty afforded by a hierarchy offers ease and effectiveness in communication.

There are, however, situations in which hierarchies can be problematic. As long as a hierarchy is "nested," where a large polygon is composed of small polygons that fit within the large one in some systematic manner, there is no loss of clarity in interpretation. When hierarchies formed on different nesting principles are superimposed, it can become impossible to extract clear meaning in the underlying data. In **Figure 7.1**, the county boundaries fit together with no overlap and cover the entire state exactly. The school districts form a nested hierarchy with smaller polygons filling larger ones with no overlap. Look within the circle superimposed on that map. The oddly-shaped school district overlaps three different counties (with rectilinear boundary lines). When these two nested hierarchies are superimposed, the meaning is lost. One cannot tell how to assign the residents of a county to a school district when there is ambiguous information! Clearly, it is important to create nested hierarchies and to do so in the broad context of considering how other related hierarchies are being created!

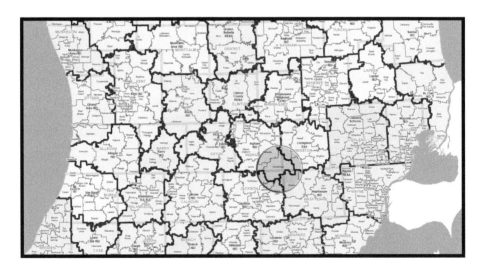

***Figure* 7.1** *County/state boundary hierarchy does not mesh with school district boundary hierarchy. One example is highlighted inside the circle. Look for others, especially in the northwest region of the map. Source: "School Districts and Intermediate School Districts Boundary Map," State of Michigan, www.michigan. gov, produced by: J. Shively, Michigan Center for Geographic Information, Department of Information Technology, Completion: March 2008. http://www. michigan.gov/documents/CGI-state_sch_district_67407_7.pdf*

7.2 Hierarchies: Census data

Governments have been collecting data, in some cases for hundreds or even thousands of years, on population, housing, agriculture, economics, and other demographic variables. The earliest documented censuses date from those undertaken in 500–499 BC in the Persian Empire (Olmstead, 1948). Sweden's 1749 census is the oldest of the modern census (LSE Library, 2012). Many countries collect data at several geographic levels, ranging from the country as a whole, to states and provinces, regions, cities, districts, and individual city blocks. Some conduct a recurring census every ten years or at other intervals, and some use the same or similar statistical areas as in the previous census years. This practice allows analysts to conduct historical comparisons. The number of questions asked, and the level of detail collected, varies considerably among the statistics agencies conducting the survey. The quality of data provided in the returns also varies, due to the mobility of the population, the manner in which the data are collected, the type of government that is collecting the data, and many other factors. As a result, some sections of the population may be underrepresented. Some census data are already mapped for the end user online, while other data are available as tables, and other are available as GIS-compatible files.

To work effectively with data, it is important to understand the hierarchy or nested relationship of how the data are collected and represented in geographic space. For example, land use data are generated from the interpretation of satellite imagery, and its resolution depends on the agency doing the interpretation and the resolution and type of the original imagery used. In the case of data from the US Census Bureau, they are collected by political area and by statistical area. The political areas include American Indian reservations, cities, counties, and states. Some political areas follow a standard: Counties nest within states, for example. But much variation exists beneath that level of geography. In most states, counties include cities. However, cities also can cross county lines. In Virginia, some cities are not a part of the county they fall within, but are independent from that county. Statistical areas, on the other hand, are not political entities but are strictly for reporting purposes. These include, but are not limited to, census tracts, block groups, census-designated places, and blocks. Blocks in an urban area are typically bounded on all sides by streets or some other human-constructed feature. Blocks in a rural area may be much larger, and may be bounded by roads, streams, or railroads. Groups of blocks are called block groups, groups of block groups are called census tracts, and census tracts nest within counties or county equivalents, such as parishes, independent cities, or boroughs. The relationships among the various statistical areas are shown in **Figure 7.2**.

In this example, County 9502 is divided into three census tracts. Each census tract is divided into block groups. Tract 1 is divided into block groups with numbers in the 100s; Tract 2 into block groups with numbers in the 200s; and

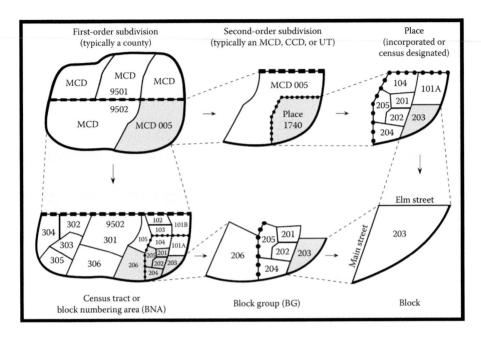

Figure 7.2 *Census Small-Area Geography. MCD stands for Minor Civil Division; CCD stands for Census County Division; UT stands for Unorganized Territory. Source: US Census Bureau, http://www.census.gov/geo/www/GARM/Ch2GARM.pdf*

Tract 3 into block groups with numbers in the 300s. The block group with numbers in the 200s is divided into blocks numbered 201, 202, 203, 204, 205, and 206. Block number 203 is focused on in the lower right. Another route to the same location, Block 203, may be found by following Minor Civil Division (MCD) data, across the top of **Figure 7.2**.

It is important to note, as shown in the added circle in **Figure 7.1**, that statistical areas may cross political area boundaries. Furthermore, while statistical areas are often bounded by physical features in the landscape; political area boundaries are frequently along imaginary lines. Census statistical area boundaries do not necessarily nest within political area boundaries, nor are they very often coincident with other data collection areas, such as precincts, election districts, recreational or tax districts, or ZIP codes.

The most important thing in working with census data is to ensure that your statistical data cover the same level of geography (such as block group) as the statistical area features you are working with. The statistical areas form the physical boundaries of your area of investigation (the *G* part of *GIS*), while the statistical data provide the demographic information for that area (the *I* part of *GIS*). If you are working with block groups, for example, make sure that the demographic data you use are at the block group level.

7.3 Thinking outside the pixel

In an earlier section, we employed a classical model, from pen and ink and paper cartography, to think about the parallels on the computer screen. Now, we stretch that style of thinking to begin with a different classical model to suggest possible futures: A model of classical central place theory reflecting an abstract characterization of how cities might share space. The interpretation here is that square pixels form a square lattice on the computer screen. One might imagine, instead, that a triangular lattice with hexagonal (or triangular) pixels could take advantage of this tightest method of packing polygons into two-dimensional space (Fejes-Toth, 1968). To consider this idea and imagine it cast in a contemporary setting, we look at some of the elements of the geometry of layers of hexagons—as tessellations that partition the plane. To see how layers of hexagons of varying size might create different pattern depending on orientation and relative cell size, consider the two different images in **Figure 7.3**. Each is based on three different layers of hexagonal cells: Large, medium, and small. The nesting patterns, while regular, are quite different from each other.

To gain an appreciation for working in a hexagonal grid, try an interactive online game based on such a grid. One on a hexagonal board with extra complexity, involving adjacency as well as jumping, may be found at: http://neave.com/hexxagon/. To view some of the intricacies of circle packing, consider the interactive graphic at: http://jsxgraph.uni-bayreuth.de/wiki/index.php/Apollonian_circle_packing. These are Apollonian circle packings.

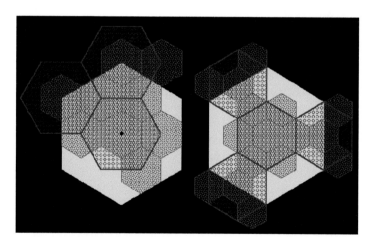

Figure 7.3 *Two different overlay patterns of three hexagonal grids, with large (green), medium (blue), and small (red) cell size. Source: Arlinghaus, S. L. and W. C. Arlinghaus. 2005. Chapter 2, Spatial Synthesis: Volume I, Centrality and Hierarchy. Book 1. Ann Arbor: Institute of Mathematical Geography. http://www-personal. umich.edu/~copyrght/image/books/Spatial%20Synthesis2/chapter2.html.*

Apollonian circle packings arise by repeatedly filling the interstices between mutually tangent circles with further tangent circles. The problems associated with close packing and related topics form a broad field within mathematics; clearly, these topics are spatial in nature.

7.3.1 Hexagonal hierarchies and close packing of the plane: Overview

Gauss (1840) proved that the densest lattice packing of the plane is the one based on the triangular lattice. In 1968 (and earlier), Fejes-Toth proved that that same packing is not only the densest lattice packing of the plane but is also the densest of all possible plane packings. If one thinks, then, of circular buffers around lattice points as if they were bubble foam, the circles centered on a square grid pattern expand and collide to form a grid of squares (Boys, 1902)—a raster. The circles centered on a triangular grid pattern expand and collide to form a mesh of regular hexagons, as do the cells in a slice of a honeycomb of bees (de Vries, 1906)—as proximity zones. The theoretical issues surrounding tiling in the plane are complex; even deeper are those issues involving packings in three-dimensional space. The reader interested in probing this topic further is referred to the References section at the end of this chapter and at the end of the book. The interpretation of the simple triangular grid has range sufficient to fill this document and far more.

7.3.2 Classical urban hexagonal hierarchies

One classical interpretation of what dots on a lattice might represent is found in the geometry of "central place theory" (Christaller, 1933, 1941; Lösch, 1954). This idea takes the complex human process of urbanization and attempts to look at it in an abstract theoretical form in order to uncover any principles which might endure despite changes over time, situation, cultural tradition, and all the various human elements that are truly the hallmarks of urbanization. Simplicity helps to reveal form: Models are not precise representations of reality. They do, however, offer a way to look at some structural elements of complexity. Thus, dots on a triangular lattice become populated places (often, villages). Circles, expanding into hexagons, are areas that are tributary to the populated places. In the traditional formulation (described after Kolars and Nystuen, 1974) one considers four basic postulates (not one of which is "real" but each of which is simple):

- The backdrop of land supports a uniform population density.
- There is a maximum distance that residents can easily penetrate into the tributary area.
- There is slow and steady population growth.
- Village residents who move, as a result of growth (or for other reasons), attempt to remain in close contact with their previous location (to maintain social or other networks).

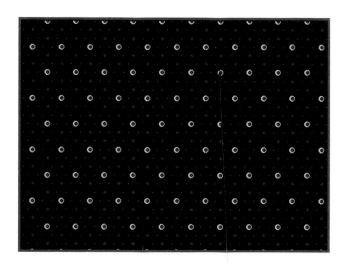

Figure 7.4 *A triangular lattice of dots with locations for competing locations entering the picture at three levels—red, the smallest, blue next, and green the largest. Source: Arlinghaus, S. L. and W. C. Arlinghaus. 2005. Spatial Synthesis: Volume I, Centrality and Hierarchy. Book 1. Ann Arbor: Institute of Mathematical Geography. http://www-personal.umich.edu/~copyrght/image/books/Spatial%20Synthesis2/ Figure1_3new.gif*

Suppose, in a triangular lattice of villages, that one village adds to its retailing activities. After some time, growth occurs elsewhere. How might other villages compete to serve the tributary areas? How will the larger, new villages share the tributary area? **Figure 7.4** shows a hierarchy of competing centers. The smallest villages are represented as small red dots; next nearest neighbors competing for intervening red dots are represented in blue; and, next nearest neighbors competing for intervening blue dots are represented in green. Of course, one is usually only willing to travel so far to go to a place only slightly larger, so the fact that the pattern could be extended to an infinite number of levels, beyond green, may not mirror the second postulate. Over time, however, one might suppose further growth and an entire hierarchy, of more than the three levels suggested here, of populated places.

7.3.3 Visualization of hexagonal hierarchies using plane geometric figures

7.3.3.1 Marketing principle

Consider a central place point, *A*, in a triangular lattice. Unit hexagons (fundamental cells) surround each of the points in the lattice and represent the small tributary area of each village. Growth at *A* has distinguished it from other villages in the system. It will now serve a tributary area larger than will the unit hexagon. There are six villages directly adjacent to *A*. The unit hexagons represent a partition of area based on even sharing of the area between *A* and

Figure 7.5 *Marketing hierarchy showing three layers of a nested hierarchy of hexagons of various sizes oriented with respect to one another according to a distance minimizing principle based on competing centers entering the system at $\sqrt{3}$ units apart—minimizing the distance to market. QR code leads to animation. Source: Arlinghaus, S. L. and W. C. Arlinghaus. 2005. Spatial Synthesis: Volume I, Centrality and Hierarchy. Book 1. Ann Arbor: Institute of Mathematical Geography. http://www-personal.umich.edu/~copyrght/image/books/Spatial%20Synthesis2/ Figure1_5new.gif*

these six villages. When *A* expands its central place activities, others may also desire to do so as well. Given that they, too, will share the area evenly, a set of larger hexagons emerges. **Figure 7.5** shows the unit hexagons and the larger hexagons based on the expansion of goods and services.

The competitors that enter are spaced at a distance, in terms of lattice points spacing one unit apart, but in terms of underlying Euclidean distance, $\sqrt{3}$ units apart. The positions of the competitors that enter the system in this scenario are as close as possible to *A*; the expansion of goods and services at any of the six closest neighbors would constitute no change in the basic pattern. One might imagine, therefore, that emphasis on distance minimization optimizes marketing capability—distance to the market is at a minimum.

Classical central place theory notes that in the marketing hierarchy, there is a principle of three involved. Competing centers are $\sqrt{3}$ units apart. Take a closer look at the hierarchy shown in **Figure 7.6**. Each blue hexagon contains

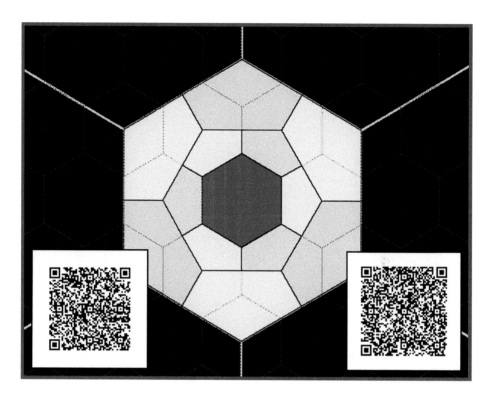

Figure 7.6 *The large green hexagon contains the equivalent of three blue hexagons (1 full one plus six 1/3 blue hexagons). Similarly, the intermediate-sized blue hexagon contains the equivalent of six red hexagons. QR codes lead to animations. Source: Arlinghaus, S. L. and W. C. Arlinghaus. 2005. Spatial Synthesis: Volume I, Centrality and Hierarchy. Book 1. Ann Arbor: Institute of Mathematical Geography. http://www-personal.umich.edu/~copyrght/image/books/Spatial%20Synthesis2/ singlek3rednew600.gif; http://www-personal.umich.edu/~copyrght/image/books/ Spatial%20Synthesis2/singlek3bluenewcropped600.gif.*

the equivalent of three red hexagons: One entire red hexagon is surrounded by six copies of 1/3 of a red hexagon. Each green hexagon contains the equivalent of three blue hexagons: One entire blue hexagon is surrounded by six copies of 1/3 of a blue hexagon. The green hexagons contain the equivalent of three blue hexagons and three-squared red hexagons. The next largest hexagon, not shown in **Figure 7.6**, would contain three green hexagons, three-squared blue hexagons, and three to the third power red hexagons. Powers of 3, fractions with 3, and the square root of three all come into play.

7.3.3.2 Transportation principle

Figure 7.7 shows the locations for the next nearest competitors, next beyond those from the marketing principle above, to enter the system. Given that they, too, will share the area evenly, a set of even larger hexagons emerges.

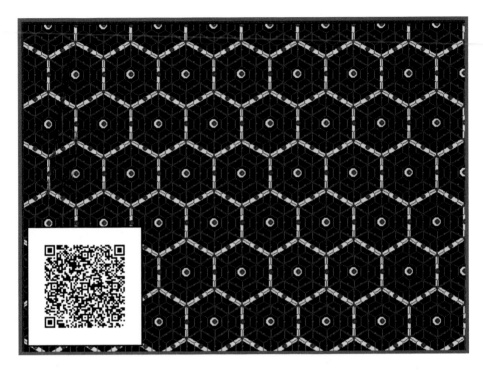

Figure 7.7 *Transportation hierarchy showing three layers of a nested hierarchy of hexagons of various sizes oriented with respect to one another according to a transportation principle based on competing centers entering the system at two (square root of four) units apart. QR code leads to animation. Source: Arlinghaus, S. L. and W. C. Arlinghaus. 2005. Spatial Synthesis: Volume I, Centrality and Hierarchy. Book 1. Ann Arbor: Institute of Mathematical Geography. http://www-personal. umich.edu/~copyrght/image/books/Spatial%20Synthesis2/Figure1_7new.gif.*

Figure 7.7 also shows the unit hexagons and the larger hexagons based on the expansion of goods and services. The competitors that enter are spaced at a distance, in terms of lattice point spacing one unit apart, but in terms of underlying Euclidean distance, two units apart. The position of the competitors that enter the system in this scenario lie along radials that fan outward from a center and pass along existing boundaries to the tributary areas. One might imagine, therefore, that emphasis on market penetration, or transportation, is the focus here. Transportation ease is minimized.

Classical central place theory notes that in the transportation hierarchy, there is a principle of four involved. The competing centers are two (square root of four) units apart. Take a closer look at the hierarchy shown in **Figure 7.8**. Each blue hexagon contains the equivalent of four red hexagons: One entire red hexagon is surrounded by six copies of 1/2 of a red hexagon. Each green hexagon contains the equivalent of four blue hexagons: One entire blue hexagon is surrounded by six copies of 1/2 of a blue hexagon. The green hexagons contain the equivalent

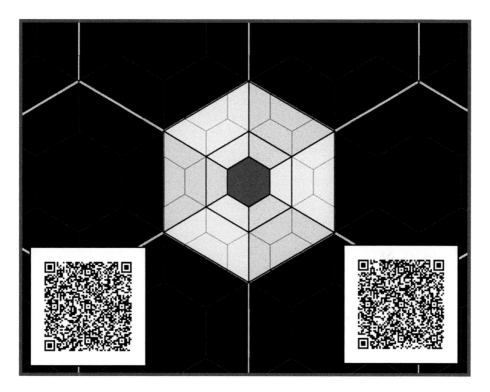

Figure 7.8 *The large green hexagon contains the equivalent of four blue hexagons (1 full one plus six 1/2 blue hexagons). Similarly, the intermediate-sized blue hexagon contains the equivalent of six red hexagons. QR codes lead to animations. Source: Arlinghaus, S. L. and W. C. Arlinghaus. 2005. Spatial Synthesis: Volume I, Centrality and Hierarchy. Book 1. Ann Arbor: Institute of Mathematical Geography. http://www-personal.umich.edu/~copyrght/image/books/Spatial%20Synthesis2/ singlek4rednewcropped600.gif; http://www-personal.umich.edu/~copyrght/image/ books/Spatial%20Synthesis2/singlek4bluenewcropped600.gif.*

of four blue hexagons and four-squared red hexagons. The next largest hexagon, not shown in **Figure 7.8**, would contain four green hexagons, four-squared blue hexagons, and four to the third power red hexagons. Powers of four, fractions with four or its root, and the square root of four all come into play.

As one might imagine hexagonal pixels creating a different form of raster, so too, one can develop in parallel various numeric observations, principles, and theorems. As shown here, some are derivative of classical central place theory. Others use more modern techniques involving fractal geometry to capture the pattern and characterize it completely. One might view this latter process as parallel to developing vector images—based on mathematical characterization. The spatial idea is the same; the mathematical tools employed are not. Rich theory on the topic is present in the References section—it may serve as a guide to uncovering directions for exciting future research.

7.3.4 Visualization of hexagonal hierarchies using mapplets

Another method of visualizing hexagonal hierarchies looks simultaneously at connection patterns between multiple hierarchical layers of urban location maps and captures them as Java Applets: As Mapplets, as was done with Varroa mite diffusion in the last chapter (Sammataro, D. et al., 1998, 2009, 2010, 2011). This process suggests a measure of visual stability of the geometric connectivity pattern that is related to the dimensions of the bounding box. It also suggests mathematical paths for developing image structure in a manner conceptually parallel to the ideas behind established vector format. **Figures 7.9a and b** show screen captures of Mapplets for the Marketing and Transportation hierarchies, respectively.

Mapplets focus on the connection patterns between successive hierarchical layers and, when K values are loaded as distances between hierarchies, they also suggest some elusive form of structural stability of the geometric form. Animated maps of central place geometry of the plane, coupled with mapplets showing animated hierarchical pattern alone, might also suggest a three-dimensional view of central place geometry (Arlinghaus and Arlinghaus, 2004). Current technology offers variety in the visualization of hierarchies, from the classical geometry of central place theory, to the practical world of census data based on the connection of the hierarchy of census units (**Figure 7.10**). In both the census and central place models, as well as in a multitude of other hierarchies, there are diagrams based on areal units with complex

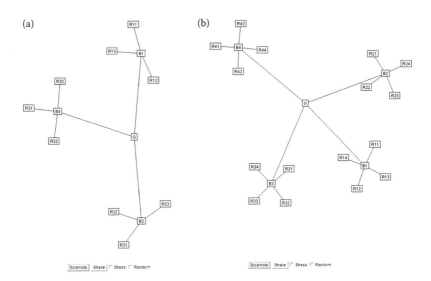

Figure 7.9 *(a) Marketing Mapplet. (b) Transportation Mapplet. Source: Arlinghaus, S. L. and W. C. Arlinghaus. 2005. Spatial Synthesis: Volume I, Centrality and Hierarchy. Book 1. http://www-personal.umich.edu/%7Ecopyrght/image/books/Spatial%20Synthesis2/*

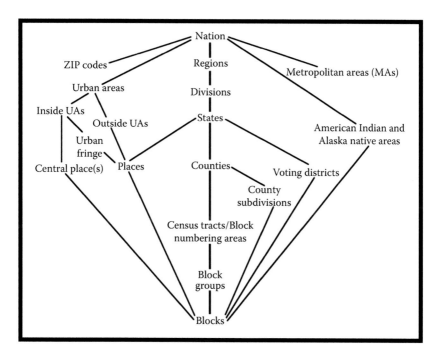

Figure 7.10 *Geographic hierarchy for the 1990 Decennial Census, represented as a structural model. Source: US Census Bureau, http://www.census.gov/geo/www/GARM/Ch2GARM.pdf*

nesting patterns (or similar relationships) and there are diagrams showing connection patterns, as mapplets or graphs, within the hierarchy.

In the following section, you will have the opportunity to use selected concepts about hierarchy from this chapter in a hands-on manner. First, you will use ArcGIS Online to examine population data at multiple scales to examine the effect of the geographic hierarchy and scale on several variables. Then, you will use ArcGIS for Desktop to use population data in an even more rigorous way through the analytical toolkit available in that software. But in both analyses, you will examine spatial pattern, hierarchy, and statistical measures.

7.4 Practice using selected concepts from this chapter

7.4.1 An introduction to census tabulation areas: Using ArcGIS online for demographic analysis

Go to http://www.arcgis.com/home and search for a map entitled "USA Demographics for Schools." Under the resulting thumbnail for the map by cfitzpatrick, open the ArcGIS Online map viewer. This map can be accessed directly by accessing http://www.arcgis.com/home/webmap/viewer.html?webmap=a59da1645fad4014b961234eb363e7b5. Click the button on the left to

Figure 7.11 *USA Population Change, 2000–2010, by State. Source of base map: Esri software.*

access the map layers, and change to the USA Population Change 2000–2010 (**Figure 7.11**).

Click the button to the left of the map to access the legend. What are the range and the mean of population change by state? The range is 0.1% growth to 3.2% growth over those 10 years, and the mean change is 0.9% growth. What pattern of population change do you notice by state? You should notice much variability, but generally, the higher growth states are in the west and south. Click on a sample of states to access the variables, the "I" part of the GIS. What state has the highest rate of population change? It is Nevada. What state has the lowest rate of population change? It is Rhode Island. What are some of the reasons for the patterns that you see? The reasons are many, but include migration to "sunbelt" states caused in part by the aging of the population and many retirees fueling this migration, agricultural mechanization and rural-to-urban migration, economies and the employment base of different regions, international migration, and other factors. How are the highest- and lowest-rate states predicted to change between 2010 and 2015? Nevada is predicted to grow at a slower rate in the near future, and Rhode Island is predicted to decline in population.

Zoom in to the county level by drawing a box with your mouse in a small area or by using the zoom controls on the side of the map. You should see how the counties nest within states. What pattern of change do you notice by county, and what are the range and the mean? The county level pattern is much more varied in pattern and also in magnitude, as evidenced by the new range of –5.5% to 9.0%, with a mean of 0.5%. Zoom in to the census tract level, choosing an area in the county that you are interested in. You should see how the census tracts nest within counties. Again, indicate the pattern of change, the range, and the mean. Your answer will vary depending on the area that you are studying, but the range should be greater than the range for the counties. Zoom in and repeat for a larger scale (that of block groups). You should see how the block groups nest within census tracts. In the map legend, the census geography you are examining (in this case, block groups) is listed at the top of the classes. Indicate the pattern, the range, and the mean. Again, your answer will vary, but the range should be even greater at the block group level. Why does the range increase as the scale becomes larger? The range increases because the more detailed the level of geography, the more variability occurs, due to the factors of age, employment, perception, and others. Why does the mean change depending on the scale? The mean changes because the mean reflects the range of its particular hierarchy of geography. Can you zoom to the block level? The reason you cannot do so is because most of these Census data items are not disseminated at the block level because of privacy issues.

Next, choose another variable in the list, that of median age, and compare patterns, ranges, and means as you change the scale. What does it signify to have a "mean" of the variable "median age?" The mean here is simply the average of the median ages in the particular level of geography that you are examining. Next, choose another variable in the list, such as the median home value or household income. Which variables seem to be positively correlated, and why? Median home value and household income seem positively correlated, as those households with more income tend to purchase and live in more expensive homes. Median age and household income also seem to be positively correlated. Which variables seem to be negatively correlated? Obviously, the median age and the population aged younger than 18 are negatively correlated, as are the percentage younger than 18 and the percentage over 64. Other variables, such as the population density and the change in population may be positively or negatively related to home value and household income, depending on the area examined. Which variables seem to have a similar spatial pattern? Again, it depends on the region examined, but home value and household income have similar spatial patterns. Make the population density layer semi-transparent and examine the topography underneath it. Population density is strongly influenced by terrain; densely settled areas tend to be flat or only moderately hilly. San Francisco is a notable exception! Finally, examine the tapestry theme and do some research on what tapestry segmentation indicates. Why can you not compare the numbers on the tapestry data in the same way as those in the other data layers? The tapestry number is a value indicating the type of lifestyle and consumer spending habits that the typical resident in that area has. It is a nominal scaled number and not a ratio scaled number. Thus, tapestry "20" is not "more" or "less" than tapestry 19.

7.4.2 Using ArcGIS desktop for demographic analysis

To analyze census data, you need two components: (1) The demographic data, and (2) the associated statistical areas as spatial data layers in GIS format. Think of the demographic data as what you "pour into" your statistical areas. One component of ArcGIS Online is US Demographics. Here we use the demographic layer from ArcGIS Online inside ArcMap. One way to access and map Census data is with ArcGIS Online, http://www.esri.com/arcgisonline. Look for where the concept of nested hierarchy enters the picture.

Here, we will use ArcGIS Online to explore racial and ethnic diversity in the USA. Start ArcMap with a new map document. Select "Add Data from ArcGIS Online" and select the USA Population by Zip Code box and click "add" to add the data as a premade map document (.mxd) to your ArcMap session (**Figure 7.12**).

Access the Bookmarks menu and select "California." Once in California, zoom to any part of the state. You will see three layers: Population Growth

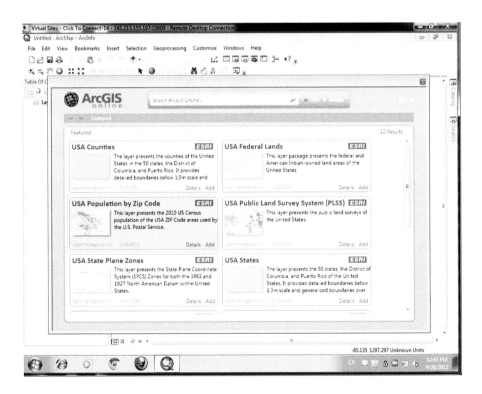

Figure 7.12 Obtain Census data online, Esri software.

2000–2008, Diversity 2008, and Median Income 2008. Turn on the Diversity 2008 layer and turn off the other two layers. The diversity index summarizes 2008 racial and ethnic diversity in the USA, showing the likelihood that two persons chosen at random from the same area belong to different race or ethnic groups. The index ranges from 0 (no diversity) to 100 (complete diversity). For example, the diversity score for the US is 60, which means there is a 60 percent probability that two people randomly chosen from the USA population would belong to different race or ethnic groups. Diversity in the US population is increasing. The states with the most diverse populations are California, New Mexico, and Texas. The geography depicts states at greater than 1:5,000,000 scale, counties at 1:1,000,000 to 1:5,000,000 scale, Census Tracts at 1:250,000 to 1:1,000,000 scale, and Census Block Groups at less than 1:250,000 scale.

The example below shows part of San Diego County (**Figure 7.13**). What patterns do you notice? Do they surprise you? How does the diversity index vary across the county? The diversity varies, but is generally higher in the southern part of the county. How does diversity range across the state of California? Again, it varies greatly, but is highest in many urban areas and along the border with Mexico. How does San Diego County compare to the county in which you live? What are some reasons for the differences? These will include past and present immigration patterns, employment, perception, the housing market and patterns, the presence of universities or other landmark

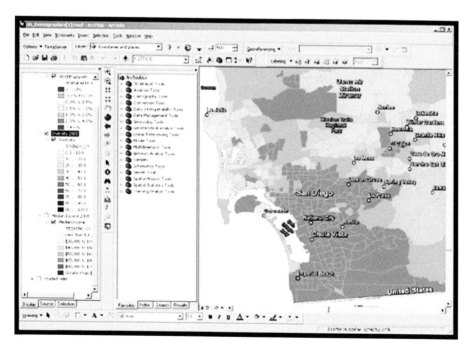

Figure 7.13 San Diego County, map made in ArcGIS, Esri.

features, the median age, and other factors. How do you think the diversity index in San Diego County or in your county has changed over time? Name some reasons why it changes. The diversity index, like other variables we are examining, changes constantly because of the reasons mentioned above that are economic and social in nature, local in scale but also affected by national and international forces.

You have just used ArcGIS Online along with ArcGIS for Desktop to create some quick maps based on Census data in the context of examining spatial patterns and analysis in order to investigate pattern of diversity in California and other places of interest to you. Investigate the other data sets in ArcGIS Online (http://www.esri.com/arcgisonline) including imagery, topography, political boundaries, land cover, and much more for your data and project needs.

7.4.3 Denver Internet café analysis

In **Chapter 3**, you created a map using a GIS called "InternetCafeSites" or some similar name using Census data as you sought to determine the ideal location for an Internet café in Denver. The activity below will build on the map you made. It will guide you through the use of Census data to create classified choropleth maps, the creation of a dot density map (foreshadowed in **Chapter 5**), and elements involving the normalization of data (foreshadowed in **Chapter 6**). A link to the full study is provided below. Again, look for where the concept of nested hierarchy enters the picture.

In **Chapter 3**, we found regions within one kilometer of a high school, university, or college, which was one of the criteria deemed advantageous to consider for the location of the Internet café. Here we will solve another component: In what Denver neighborhoods is the percentage of 18 to 21 year-olds greater than 10% of the total population of those neighborhoods? This demographic characteristic is also deemed useful for the café's location. These neighborhoods tend to be near universities, such as in northwest Denver, in downtown Denver, and in south central Denver, the sites of Regis University, the University of Colorado at Denver, and the University of Denver, respectively.

Open the table for the Block Groups polygons for Denver City and County. How many records exist in this table; that is, how many block groups exist in Denver City and County? There are 470 block groups in Denver City and County. What geographic extent do you think the longer table of Census variables covers? The longer table, the one with 3278 records, is all of the block groups for the state of Colorado. What is the name of the common field that these two tables share? The common field is named STFID, or "state federal ID."

Notice that your map file (shp file) contains block groups but no Census data. Your data table contains Census data but no map. In the next step, you will join the data table to the map. Clear any selected records in both tables so that you have no records selected. A "join" in GIS usually operates on only the selected records, and if none are selected, the join works on all of the records. Therefore, it is important not to have any selected before the join is performed. Right click on the tgr08031grp00 map layer in the table of contents and use the "Joins and Relates" function. Join *from* the data table *to* the map attribute table so you can map the demographic data for Denver County block groups. Refer to ArcGIS Help if you need to. After joining your tables, right click on your tgr08031grp00 map layer again, and using Data→ Export Data, export your joined data to a permanent data file named DenverCountyBlockGroups. Save your ArcMap document.

For your new DenverCountyBlockGroups layer, symbolize it as a graduated color thematic map of population 18 to 21 years old. Next, symbolize the block groups as a graduated color thematic map of population 18 to 21 years old normalized by the 2000 population of the block group.

What does normalizing the data do? Normalizing divides one variable by another; in this case, normalizing in effect creates a map showing 18 to 21 year-olds as a percentage of the total population of the block group.

Describe the resulting geographic pattern of percentage of 18 to 21 year-olds by block group in Denver. The pattern shows generally low percentages of 18 to 21 year-olds, except for the west side of Denver, and near the three university campuses. It will be much easier to solve your problem if you create a field that contains the actual percentage of 18 to 21 year-olds by block group. To do this, you will need to add a field and calculate the value of the new field that represents the percentage 18 to 21 year-olds in each block group. Name the new field p1821, make it a float (floating point) data type, and give it a precision (width) of 5 and a scale (number of decimal places) of 2.

You may receive a VBA (Visual Basic for Applications) overflow error. If so, cancel the calculation and examine the table at the row where the percentage vacant field stopped calculating. Why did the calculation stop working at the point where it did? Hint: Is it possible to divide by zero? The calculation may stop when it encounters a division by zero. One of the block groups has zero population (an industrial or military block group), and at this point where 0 was in the denominator, the calculation stopped. How can you select certain records in the table so this problem will not occur again? You can select block groups where the total population was greater than zero. After selecting only those records where the block group population is greater than zero, recalculate.

Make a thematic map of your new field p1821. Indicate the sections of Denver where the highest percentage of 18 to 21 year-olds are found. Next, make a pie chart map where the two parts of each pie represent the following: In the first part, the percentage of 18 to 21 year-olds, and in the second part, the total population in each block group. Experiment with the size and the overlap functions.

Next, make a dot density map of the raw numbers of 18 to 21 year-olds. In your opinion, which of the maps—graduated color, pie chart, or dot map—is the easiest to understand for mapping this type of variable at a county-wide scale, and why? The disadvantage of the pie chart map at the scale of examining the whole of Denver County is that there are so many of them. But at larger scales, the pie chart map might be the ideal map to use because you can vary what each wedge shows, thus increasing the amount of understandable content in a small space. How many block groups exist that have 10% or more of their population between 18 and 21 years of age? You have now determined another criterion for locating an Internet café in Denver.

7.5 Related theory and practice: Access through QR codes

Theory

Persistent archive:

University of Michigan Library Deep Blue: http://deepblue.lib.umich.edu/handle/2027.42/58219

From Institute of Mathematical Geography site: http://www.imagenet.org/

Arlinghaus, S. L. and M. Batty. 2006. Visualizing Rank and Size of Cities and Towns. Part II: Greater London, 1901–2001. *Solstice: An Electronic Journal of Geography and Mathematics*, Volume XVII, Number 2, Institute of Mathematical Geography. http://www-personal.umich.edu/~copyrght/image/solstice/win06/arlbat2/indexPartII.html

Arlinghaus, S. L. and W. C. Arlinghaus. 2005. *Spatial Synthesis: Volume I, Centrality and Hierarchy. Book 1*. Ann Arbor: Institute of Mathematical Geography. http://www-personal.umich.edu/~copyrght/image/books/Spatial%20 Synthesis2/

Arlinghaus, S. L. and W. C. Arlinghaus. 2004. Spatial Synthesis Sampler. Geometric Visualization of Hexagonal Hierarchies: Animation and Virtual Reality. *Solstice: An Electronic Journal of Geography and Mathematics*. Volume XV, No. 1. Ann Arbor: Institute of Mathematical Geography. http://www-personal.umich.edu/~copyrght/image/solstice/sum04/ sampler/index.html

Arlinghaus, S. 1993. Microcell Hex-nets. *Solstice: An Electronic Journal of Geography and Mathematics*. Volume IV, No. 1. Ann Arbor: Institute of Mathematical Geography. http://www-personal.umich.edu/%7Ecopyrght/image/solstice/ sols193.html

Arlinghaus, S. 1990. Fractal geometry of infinite pixel sequences: 'Super-definition' resolution? *Solstice: An Electronic Journal of Geography and Mathematics*. Volume I, Number 2. Ann Arbor: Institute of Mathematical Geography. http:// www-personal.umich.edu/%7Ecopyrght/image/solstice/sols290.html

Practice

From Esri site: http://edcommunity.esri.com/arclessons/arclessons.cfm

Kerski, J. Siting an Internet Café in Denver Using GIS. Complexity level: 3. http://www.mylovedone.com/Kerski/Denver.pdf

Kerski, J. Siting a Wind Turbine on Your School Campus. Complexity level: 3. http://edcommunity.esri.com/arclessons/lesson.cfm?id=603

Kerski, J. Analyzing Current Wind Speed and Direction in North America. Complexity level: 3. http://edcommunity.esri.com/arclessons/lesson.cfm?id=600

Kerski, J. Siting a Wind Farm in Indiana. Complexity level: 3. http://edcommunity.esri.com/arclessons/lesson.cfm?id=601

Kerski, J. Exploring the San Gorgonio Wind Farm, California. Complexity level: 3. http://edcommunity.esri.com/arclessons/lesson.cfm?id=602

Kerski, J. Acquisition, Integration, Mapping of Census Data. Complexity level: 4. http://edcommunity.esri.com/arclessons/lesson.cfm?id=439

Distribution of Data
Selected Concepts

Keywords: Tornado(es) (78), data (43), mean (36), center (31), map (22)

Be Prepared!	Beware, the Ides of March
Boy Scouts of America	**William Shakespeare**

8.1 Introduction

Dateline: Ann Arbor, Michigan; March 15, 2012.

An F3 tornado hits 110 homes, 10 miles to the northwest of the City of Ann Arbor (in Dexter, Michigan), right around dinner time. There was extensive property damage. Yet no one was killed or even injured; citizen response, at a time of day when many were likely to have been in their homes having dinner, was outstanding. The Washtenaw County website credits fine county-level planning of an emergency services network that warned the public to take shelter. Evidently, there had also been sufficient education of the public, in advance, to create an appropriate response (Washtenaw County website, 2012).

> Following the tornado that hit Dexter, Sheriff Jerry Clayton with Washtenaw County credited the preparedness of the Dexter community for saving the lives of everyone hit by the tornado. Sheriff Clayton was quoted saying, "We think that's a testament to the emergency warning system, a testament to public education as to how you respond, and a testament to the Dexter residents." As with all severe weather events, being prepared is key to keeping our communities and residents safe.

The problem of creating an emergency services and warning network is rooted in spatial and mathematical concepts. Human beings can be considered scattered as "data points" across a region. The pattern is irregular and, to some extent, unknown. At best, it is imprecise: We may have street addresses associated with individuals but we do not know where an individual is at any given time, and even if at home, where he or she is located within the home. What can be pinpointed is the location of warning stations to distribute information to the scattered population. The problem is how to locate the siren towers so as to effectively communicate to all in the coverage target with no gaps in communication.

In the material below, we consider these problems in the context of the tornado theme. However, we encourage the reader to think more broadly about distributions of data—what might one wish to know about scattered data, independent of context? To select the concepts and activities from the vast array available, we let the tornado study help to guide our selection. Here are a few ideas; we hope the reader will add more.

- Is the pattern of data regular, as in the case of the model of an urban hierarchy in the last chapter, or is it irregular? Why does it matter? Is irregularity less predictable than regularity?
- How are the data points clustered? Are they clustered together in space—trailer parks have tighter clusters of homes than do estates on large parcels. Think about the importance of clusters of housing, or their lack of clustering, in terms of tornado damage or other phenomena.
- How might clusters of data be captured for analysis purposes? All data within an ellipse of a certain size? All data within a certain distance of a central point?
- Are clusters, or lines (tracks), or other patterns of data constant over time? Are there seasonal variations in regularity or irregularity of data patterns?

Think broadly about ideas such as these and interpret them in the context of tornado activity. One might view them as "hubs" of communication (Campbell and O'Kelly, 2012) or adopt other interpretive viewpoints. Then, participate in the accompanying activity that focuses on one aspect of the analysis of clustering of data using the concepts of mean center and standard deviational ellipse. There are many other topics to consider in association with the distribution of data in time and space. What matters is to take a thoughtful approach to the selection of topics offered here and to related ones that are available elsewhere.

8.2 Ann Arbor, Michigan—Tornado siren infill project[*]

One way to get information about how effective communication might be is to listen to concerned members of an educated public. A number of years ago, the Director of Environment Coordination Services for the City of Ann Arbor, Matthew Naud, noted that members of the public had been calling him to say they could not hear the test warnings from the emergency siren warning network scheduled on Tuesdays. After mechanical testing of equipment and field checking of complaints by municipal staff, the locations of the callers were mapped in Geographic Information Systems software. Then, it became a straightforward matter to suggest possible locations to the City, for their consideration, for two new sirens to fill coverage gaps that the public had identified. The strategy below outlines the process actually employed in this data distribution problem. The reader will have other opportunities in the "Practice" section at the end of this chapter.

8.2.1 Filling gaps in tornado siren coverage: Ann Arbor, MI

Following extensive field work by the City of Ann Arbor, simple location by street position of existing towers followed by a tessellation by Thiessen polygons of that distribution, offered simultaneous display on a map of the existing siren locations partitioned according to the proximity to siren. In **Figure 8.1**, the red dots show the existing siren location and the heavy red lines show the Thiessen tessellation of those dots. What the tessellation alone does not show is how far away any given point, within the Ann Arbor freeway ring, is from an existing siren.

To get an added visual fix on this concept, concentric ring buffers were introduced around each dot. Successive buffers have radii of 1000, 2000, 3000, 4000, and 5000 feet. This pattern conveys the fact that sound and hearing

[*] With many thanks and greatest appreciation for input, in the early 2000s, from: Matthew Naud, Environmental Coordination Services, Director, City of Ann Arbor; Merle Johnson, Information Technology Services, City of Ann Arbor; Adele El Ayoubi, Neighborhood Watch Coordinator, City of Ann Arbor Police Department; Karen Hart, Planning Director, City of Ann Arbor.

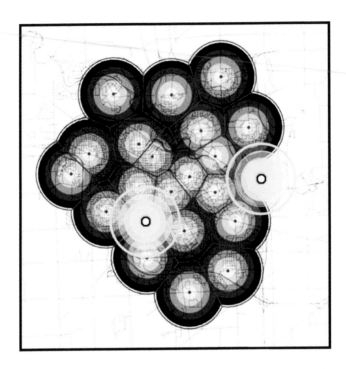

Figure 8.1 *Cyan concentric circles target locations for two new tornado sirens, represented as yellow dots. These two new locations complete the coverage for Ann Arbor. Source: Arlinghaus, S. L. 2003. Tornado Siren Location: Ann Arbor, Michigan. Solstice: An Electronic Journal of Geography and Mathematics. Volume XIV, No. 1. Source of data; base map: City of Ann Arbor; Esri software.*

decay over distance. Scale is local enough that the curvature of the Earth, and associated distortion of buffers or Thiessen polygons, may be viewed as flat. Thus, buffers of points appear as circles. The boundaries generated by these intersecting circular buffers (mostly at the 5000 foot distance) are precisely the edges of the Thiessen tessellation, as we saw in geometric constructions in **Chapter 3**. Thus, the buffers fit exactly within the Thiessen tessellation and serve as a proximity contouring of the urban surface.

What became clear from the map was that almost everyone was within 5000 feet of a siren; existing local wisdom at the time suggested that one mile was the outer edge for siren audibility. Two exceptions were evident: One southwest of downtown near The University of Michigan golf course and one at the eastern edge of the city. The position for a new siren near the golf course was easy to determine; the uncovered area was small and there were very few choices for position. The location for a new siren in the broader area to the east was found by digitizing the boundary of the uncovered area, calculating the centroid of the digitized region, and then using the centroid as the proposed siren location. In constructing the actual siren, centroid position gave a rough guide to the siren location on which to superimpose considerations such as property rights, ease of siren maintenance, access, and security. The City installed two new towers and sirens at these locations (yellow and cyan configuration in **Figure 8.1**) to complete the coverage so that all residents were now within 5000 feet of a siren.

8.2.2 Related research

A number of open abstract questions remain: Are the suggested locations for new sirens the best locations for new sirens? What sorts of analytical tests might be employed to see how well the suggested locations fit the existing distribution of data? See the outstanding work of Morton O'Kelley for a careful discussion of such matters (Campbell and O'Kelly, 2012, for example). Often, there is a gap between the scholarly analysis of real-world problems and the implementation of solutions. Sometimes the "best" solution is one that is simple, easy to communicate and to implement, instead of the one that is technically "best" in some manner. Of course, the optimum is to have the two approaches mesh; however, that does not always happen.

8.3 Educational and marketing efforts to the public

A final stage in this planning process to fill gaps in an emergency notification network was to raise public awareness. Some earlier work during the late 1970s involving the location of an emergency telephone network on The Ohio State University campus generated some local interest there, through some material that appeared in the *Columbus Dispatch* (Lease, 1979). In Ann Arbor as well, local municipal officials and others were very effective in getting media coverage recognition from a variety of directions. It appears that emergency network location problems are an attractive subject to the media! A summary appears below, at the end in the References section.

By 2003, careful planning for emergency notifications, and associated public education, had been implemented by outstanding municipal authorities in the City of Ann Arbor (Washtenaw County, Michigan). In 2012, the public were able to see, in a different way, that County-wide planning on the part of County municipal officials was at least as effective as in the City. They were also able to see how spatial analysis solved a real world problem. Unfortunately, the Ides of March test of the county-wide system came about in the real-world lab of a violent tornado and of an associated storm entering the system.

Following the tornado, local community activist and mapping enthusiast Roger Rayle mapped the path of this one tornado in Google Earth and then set it within the broader spatial context of tornadoes in Michigan. His map (**Figure 8.2**), suggests broader uses for mapping software in association with tornado tracking and with mapping other phenomena that have both a point

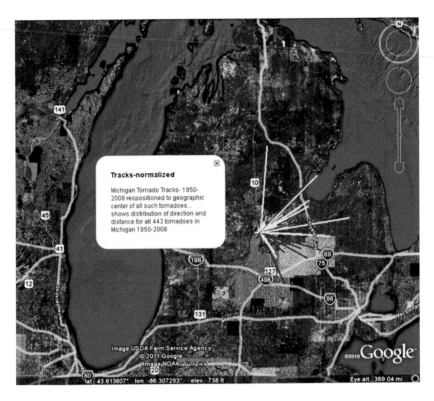

Figure 8.2 Tornado tracks in Michigan. Source: Roger Rayle gave permission to use his image of re-rendered data originally from, a Google Earth basemap at http:// www.srh.noaa.gov/gis/kml/TIMS/USTornadoes1950-2008-TIMS.kml.

component and a vector component (see activities in the Practice section of this chapter).

In the next section, you will have the opportunity to examine tornado data yourself, using the data produced from a web-based GIS, while considering concepts of data quality, direction, length, temporal patterns, spatial patterns, density, and other factors.

8.4 Examining the distribution of tornado data

In this section, you will consider key mathematical and spatial concepts while using an online GIS to examine tornadoes. Consider the distribution of tornadoes in the USA as gathered by the National Oceanic and Atmospheric Administration (**Figure 8.3**).

This map layer shows the tornado touchdown points in the United States, Puerto Rico, and the US Virgin Islands, from 1950 to 2004. Statistical data were obtained from the National Weather Service, Storm Prediction Center

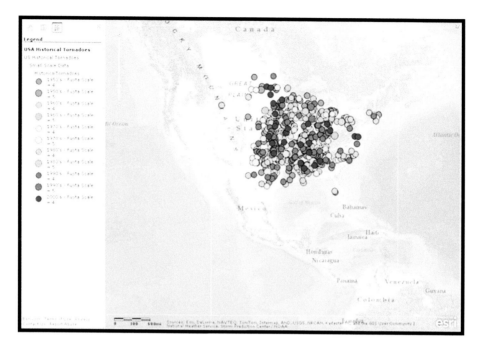

Figure 8.3 *Tornado touchdown distribution colored by Fujita scale, 1950–2004. Source of base map: Esri software.*

(SPC). The SPC data originate from the Severe Thunderstorm Database and the National Oceanic and Atmospheric Administration (NOAA) Storm Data publication. This map shows the distribution of tornadoes of Fujita (intensity) Scale 4 and 5. The predominant pattern of tornadoes in the central and eastern United States is evident, with gaps in the Appalachians, the Gulf and Atlantic Coasts, and New England. At this scale, it does not appear that any tornadoes occur in the mountain west. But upon zooming into a larger scale, the scale-dependent map shows Fujita 2 and 3 tornadoes, and it becomes evident that mountain west areas, such as Colorado, that have not been visited by a higher-intensity tornado have been visited by a lower-intensity tornado, not only once, but numerous times (**Figure 8.4**).

This map provides an excellent illustration of the issues of spatial data quality that we have been discussing throughout this book. Doppler radar did not come into widespread use until the 1980s. Most of the tornado touchdown data points from the 1950s through the 1970s, therefore, were placed where they are as a result of local observations and damage reports. Thus, those locations are approximate and based on local stories. Furthermore, if a tornado occurred far from a road or house where people could observe it, chances are that it is not in the data set. Therefore, can you detect any predisposition of tornadoes from the 1950s through the 1970s to be close to roads, towns, and other locations where people are typically found?

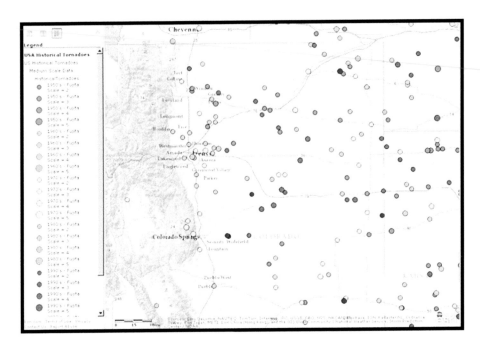

Figure 8.4 *Tornadoes in Colorado—lower on the Fujita scale. Source of base map: Esri software.*

What areas of the country experience the most frequent tornadoes? In other words, as manifested on the map, which areas of the country have the densest pattern of tornadoes? Does it change any of your stereotypes about tornadoes? Contrary to popular opinion, Kansas does not appear to be the primary candidate for the most frequent and densest tornado activity, but rather, Oklahoma.

Tornadoes are not damaging simply because they touch down, but in large part because they move across the landscape, tearing up land and infrastructure as they do so. Therefore to fully understand the phenomenon spatially, it is necessary to examine tornadoes as tracks and not as points. For example, examine the map in **Figure 8.5**, showing tornadoes mapped as tracks where the track data are known, with an arrow indicating the direction of movement. What spatial patterns do you notice to the length and direction of the tracks? You should detect a southwest-to-northeast movement of most of these tornadoes, no matter where in the country you are looking. Numerous exceptions exist, but this direction fits with the prevailing wind direction and the jet stream across North America. Measuring the mean direction of the tornadoes on this map, in a 25 mile radius around Ann Arbor by cardinal direction, reveals a direction of approximately 60, or 60 degrees east of north, which is a northeast-tending-to-due-east direction. An analysis of the track length of these tornadoes reveals the mean track length as eight miles (12.9 kilometers).

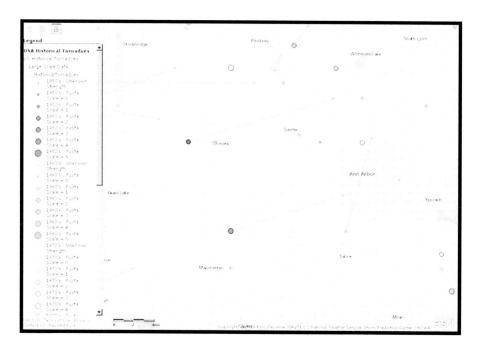

Figure 8.5 *Mean tornado tracks, Ann Arbor, MI. Source of base map: Esri software.*

Yet even though we are now examining tornadoes as tracks and not simply as point data, do you think the tracks are entirely accurate? Do you think they are actually as linear as they appear on the map, or can they be sinuous in shape? You should be thinking about the influence of local terrain and the rapidly changing nature of storms that could result in a sinuous path, or even a tornado that touched down and lifted up numerous times during its path. Also, do you believe that the track of destruction is as narrow as it appears as represented by a single line? Do some research on tornado width. What is the typical range of width of tornadoes? You will discover that tornadoes range from a few meters to hundreds or even thousands of meters (yes, several kilometers) wide. How could this width be portrayed on a map? This width could be shown as a gray or hatch pattern in a polygon, instead of a line. But the limitations there are that data on the width of tornadoes as they move is extremely difficult to measure, given their short life spans and the difficulty of obtaining such measurements.

Furthermore, consider that besides the destruction caused *within* the tornado, much destruction occurs *near* a tornado. In other words, tornadoes wreak havoc far beyond their own track lengths and widths as portrayed on maps like this. How could this proximity be represented on the map? Again, the zone of destruction could be shown as a polygon with shading that becomes lighter farther away from the tornado and darker near the tornado. Recall your earlier work with buffers. Applying a standard buffer around a tornado to define the polygon could be problematic since tornadoes vary widely in wind speed, width, and track length. Adding further complexity to the accurate mapping of tornadoes is their frequent tendency to touch down, lift up, and touch down again, sometimes multiple times in the same storm.

In the next activity, you have the opportunity to make maps of tornadoes yourself, using the tools that produced the maps that you have just examined.

8.5 Activity: Examining the distribution of tornado data

Now it is your turn to investigate tornadoes using a web-based GIS, ArcGIS Online. Access ArcGIS Online (http://www.arcgis.com/home) and start a new map. Use the Add dropdown menu to add the map layer "USA Historical Tornadoes" from ArcGIS Online. Select "Done Adding Layers" and click on the button that allows you to view the map legend. What spatial patterns do you notice? You should notice that the higher-intensity tornadoes shown at the scale of the entire country predominantly occur in the central and southern states. Zoom and pan the map and notice the pattern of the tornadoes spatially and by intensity. What influence does topography, or altitude, especially in Colorado and Wyoming, seem to have on tornado frequency and intensity? Why? Mountains and hills seem to have a negative influence on tornadoes, because they inhibit a smooth track for the tornadoes to travel over and, more important, they inhibit the merging of the cool and warm air masses and the vertical updrafts and downdrafts in the storms themselves. What influence does latitude have on tornado frequency and intensity? Why? Latitude has a negative influence on tornado frequency because it is more difficult for the necessary warm, moist air masses to penetrate higher latitudes.

Click on individual tornadoes near the Gulf of Mexico and observe the month in which they occurred. Repeat this process for tornadoes in Minnesota and North Dakota. What hypothesis could you make about how the tornadoes vary spatially by month of the year? Because of latitude, weather patterns, and the proximity to oceans, January and February tornadoes tend to occur near the warm waters of the Gulf of Mexico and the Atlantic Ocean off of the coasts of Alabama, Mississippi, and Florida. They increase in frequency with the advent of spring, and also their total spatial distribution expands. They reach a numeric maximum in April when the difference between the cold air masses from Canada and the warm air masses from the Gulf of Mexico are at their greatest difference. However, they do not reach a spatial maximum until July, when most of the continent has warmed up sufficiently to support the merging of the air masses and the updrafts and downdrafts necessary for the widespread occurrence of tornadoes. Zoom to specific large cities. Can you find instances where tornadoes moved through the downtown sections of large cities? Examples include Nashville and Oklahoma City. What safety hazards are acute in such situations? These hazards include damage to infrastructure such as buildings, highways, and power lines, but more important, the flying debris from such damage can inflict further damage far from their source.

Consider the discussion earlier in this chapter about the density of tornadoes. Pan and zoom the map to be centered on Goodland, Kansas. What would you estimate the density to be in terms of tornadoes over the time period indicated per 100 square miles (an area 10 miles × 10 miles)? What is the density per square mile? If the number of tornadoes in your chosen 100 square mile area is four tornadoes, the density is one tornado per 25 square miles. What is the likelihood of a tornado in a single year in that 100 square mile area? Since there were four tornadoes in 100 square miles over a 50-year time span, the likelihood of a tornado in any single year is $4/50 = 0.08$, or 8%. Now, pan and zoom the map to be centered on Oklahoma City. What would you estimate the density to be in terms of tornadoes over the time period indicated per 100 square miles (an area 10 miles × 10 miles)? Oklahoma City seems to have had about 20 tornadoes in 100 square miles. What is the density per square mile? The density is 20/100, or 1/5, or one tornado every five square miles. What is the likelihood of a tornado in a single year in that 100 square mile area? The likelihood of a tornado in a single year in that 100 square mile area that includes Oklahoma City is 20/50, or 0.4, or 40%. Compare your Kansas and Oklahoma measurements. This 40% likelihood is five times greater than the sampled Kansas likelihood of 8%. Is there a location elsewhere in the USA where the density is higher than central Oklahoma? Central Oklahoma appears to be the area with the highest tornado density, but test areas in Texas, Missouri, and Illinois as well, since these areas seem to be prime spots for tornadoes.

We have been examining the distribution of data through attributes, proximity, density, and showing data as points, lines, and polygons, while touching on the concepts of scale and data quality. In the next section, we will turn our attention to other measures of distribution of data, including measures of centrality.

8.6 Mean center and standard deviational ellipse

Many measures of spatial distributions exist that touch on the disciplines of geography as well as mathematics. Buffering, as was discussed earlier in this work, and tessellations, discussed in the last chapter, are two such measures. Space in this book does not permit due treatment of all these measures, but two measures that are powerful and yet easy to understand are the concepts of a mean center and a standard deviational ellipse. A mean geographic center is the point representing the average x and y values for the input feature centroids. The Mean Center is given as

$$\bar{x} = \frac{\sum_{i=1}^{n} x_i}{n}, \quad \bar{y} = \frac{\sum_{i=1}^{n} y_i}{n}$$

where x_i and y_i are the coordinates for feature i, and n is equal to the total number of features. The Weighted Mean Center becomes

$$\bar{x}_w = \frac{\sum_{i=1}^{n} w_i x_i}{\sum_{i=1}^{n} w_i}, \quad \bar{y}_w = \frac{\sum_{i=1}^{n} w_i y_i}{\sum_{i=1}^{n} w_i}$$

where w_i is the weight at feature i.

For example, the mean geographic center for the state of Michigan could be constructed from the centroids, or the mean centers, of each of the state's 83 counties. The mean center can be thought of as the point at the tip of a pencil that "balances" whatever data are on a "plane" surface. A weighted mean center "weights" the geographic mean center by influencing its location based on the spatial distribution of some phenomenon, such as population, sirens, or water wells, for example. If that variable is more numerous on one side of the plane, the pencil point, or mean center, must be moved to accommodate the extra weight on one side of the plane to keep the plane in balance.

Another way to represent the distribution of data is through a directional distribution, or a standard deviational ellipse. This ellipse wonderfully summarizes the spatial characteristics of geographic features, including central tendency, dispersion, and directional trends. This ellipse is created by calculating the standard distance, or the distance within one standard deviation of the mean center of a set of data, first from the x-coordinates, and then from the y-coordinates; in other words, separately in the x-direction and then in the y-direction. These two measures define the axes of the ellipse. The size of the ellipse indicates how spatially diffuse or concentrated a set of data are spatially, and if the distribution of features is elongated, or oriented in a particular direction, that too will be evident in the standard deviational ellipse. For example, a specific disease such as schistosomiasis in the Philippines may display a standard deviational ellipse that reflects the wetlands and the vulnerable populations near those wetlands, whereas the same type of ellipse for Acquired Immuno Deficiency Syndrome (AIDS) might be larger in shape and less elliptical, reflecting the broader range of affected people in rural and urban areas.

Mean centers, weighted mean centers, and standard deviational ellipses make excellent instructional tools. A common example shown in most geography textbooks is the mean center of the population in the USA from 1790 to 2010. Over those 220 years, this mean center moved from Maryland to southwestern Missouri, reflecting the migration and the settlement patterns diffusing from the east coast of the country to the west and, later, to the south. A population mean center for Michigan lies to the south and east of the geographic center, reflecting the larger population in cities such as Lansing and Detroit, which lie in the southern part of the state. These population centers and higher rural population density surrounding them "pull" the mean center of population away from the more sparsely populated northern sections of the state. However, a mean center of hard rock mines in Michigan would lie to the north of the geographic center, reflecting the historic concentration of iron and copper mining in the Upper Peninsula in the northern part of the state.

Throughout this book, we have emphasized the care that must be taken in using and interpreting the results of spatial analysis. In the case of mean centers and standard deviational ellipses, the input polygons and therefore the centroids of those polygons is important. Changing the input data alters the position of the mean center. For example, if cities are used instead of counties in computing the mean center of population for Michigan, the mean center will be in a different location. Another important influence on the location of the mean center and standard deviational ellipse is the map projection that is chosen. Since the map projection affects the input and final position of the mean center, an equal area projection is best.

In the next section, you have the opportunity to compute the mean center and standard deviational ellipse yourself using a GIS.

8.7 Activities using mean center and standard deviational ellipse

8.7.1 Computing and analyzing mean center and standard deviational ellipse using historical population data

In the activity earlier in this chapter using ArcGIS Online, you used a web-based GIS to visualize spatial and temporal patterns of tornadoes. However, the online platform lacked the statistical tools that allow for more rigorous quantitative analysis. In the following activity, you will use ArcGIS for Desktop, with its extensive geoprocessing and spatial statistics tools, to more thoroughly investigate a data set. The data you will use is the list of the world's 10 most populous cities at selected periods over the past 2000 years. You will examine the pattern of the 10 most populous cities from the year 100 to the year 2005, noting the changes in spatial pattern and populations. Next, you will predict the 10 most populous cities in 2050 based on the growth rates of countries around the world. You will think critically about how population figures were derived in ancient times and today.

To begin, download the data from the ArcLessons library, on: http://edcommunity.esri.com/arclessons/lesson.cfm?id=411. Open the .mxd (ArcMap GIS project file, a map document) that is contained within the zip file. A map showing the 10 most populous cities for each year from the year 100 to 2005 will be visible. Next, examine the geographic mean centers of the 10 most populous cities for each year. The center represents the pencil point that would balance the 10 most populous cities if the cities lay on a plane surface. The mean centers are weighted by population, so the larger cities pull at the mean center more than the smaller ones (**Figure 8.6**).

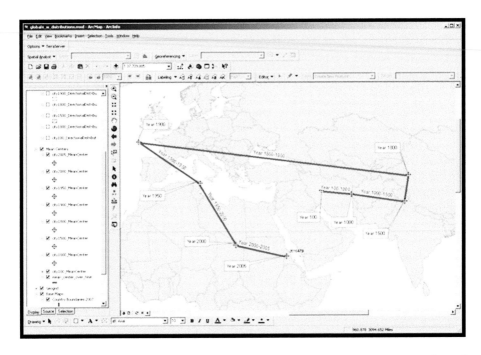

Figure 8.6 *Mean center of population for the 10 most populous cities, 100 through 2000. Source of base map: Esri software.*

The mean weighted center for the 10 most populous cities has moved counterclockwise over the past 2000 years, starting in Iran and landing in Eritrea. Describe and explain the direction and amount of movement of the mean center for each time period indicated. The mean center runs in a counterclockwise motion, heading east from the Middle East into central Asia, and then northwest to western Europe by 1900, and then heading south and east again to today. Why do you think the mean population center was off the coast of western Europe in 1900? There are many reasons, but foremost is the rise of the Industrial Revolution and the resulting urbanization of western and central Europe, with cities such as Manchester, London, Paris, and Berlin all dramatically increasing in size. Also, the rise of cities in North America such as New York City and Chicago also pulled the mean center towards the west. Examine the tables of data. What were the populations of the largest cities in 1900 and where were they located? In 1900, the largest cities were London, New York City, Paris, and Berlin, in that order. Predict where the mean center will be by 2050 given the data at your fingertips, including the growth rate by country and other variables. Many scenarios are possible, but given the current high growth trends of several countries in Africa and the Middle East, the mean center could be offshore, south of Iran perhaps, reflecting the high populations of India and also of central African countries such as Uganda, and also the continuing pull from the high population of China and southeast Asian countries such as Indonesia and Vietnam.

8.7.2 Standard deviational ellipse

Next, examine the directional distribution for the top 10 cities for each of the years under study. This standard deviational ellipse measures whether a distribution of features exhibits a directional trend (**Figure 8.7**). The smaller the ellipse, the more clustered the top 10 cities are. If the spatial pattern of the features is concentrated in the center with fewer cities toward the edges, a one standard deviation ellipse covers 68% of the cities. When was the ellipse the smallest, and why? The ellipse was smallest in 1900 as industrialization attracted migrants to and concentrated people in large cities such as London, Paris, and New York City. What do you think the standard deviational ellipse will look like in 2050 given the current population trends? Many scenarios are possible, but it is likely that given the increasing population in Brazil, Nigeria, the Middle East, and India, the standard deviational ellipse will cover an even greater portion of the Earth than today. It will also tend toward the south because northern countries in Europe and also Russia experience slow growth and in some cases, population decline.

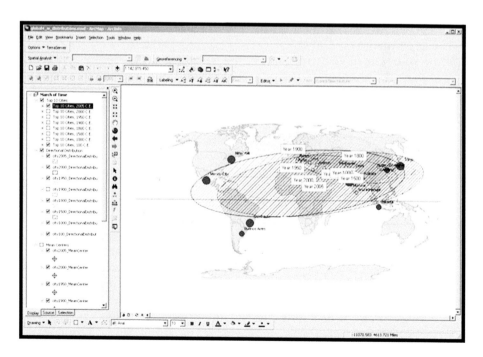

Figure 8.7 *Standard deviational ellipse. Source of base map: Esri software.*

8.7.3 Applying measures of distribution to tornado data

Earlier in this chapter, you used ArcGIS Online to examine the spatial and temporal distribution of tornadoes in the USA. Now, you can apply your new skills of measuring distribution to the same type of tornado data you were examining earlier. To do this, you will need to download the tornado data and examine it in ArcGIS for Desktop. First visit the ArcLessons library to obtain the data: http://edcommunity.esri.com/arclessons/lesson.cfm?id=565. Download and unzip the data, and open the MXD (map document) in ArcGIS for Desktop. Perform the following tasks:

1. Analyze the tornado attribute table. How many tornadoes are in this data set, and what is the range in years for this data set? The range is 1950 through 2004, and there are 46,931 tornadoes in this data set.
2. Where and when did the tornado that caused the most injuries occur? Where and when did the tornado that caused the most fatalities occur? The most injurious tornado occurred in northern Texas in 1979, injuring 1740 people, while the most fatal tornado occurred in Michigan in 1953, killing 116 people.
3. Perform a spatial join to join the tornado points to the state boundaries. To do this, right click on the tornado touchdowns data layer, use Joins and Relates, and choose "join data from another layer based on spatial location." In the dialog box, join states to the tornado points, saving your results in a shapefile named "tornadoes_with_state_names.shp." Following the join, a state name will be associated with each tornado.
4. In your joined data set, access the attribute table for the tornado points. Right click on the state_name field and select "summarize" to obtain the number of tornadoes by state. Place your result in a .dbf table with a logical name, such as "tornadoes_by_state.dbf." Sort this new table on the number of tornadoes in each state. Which state contained the most tornadoes according to this data set? Texas by far experienced the most tornadoes, followed by Oklahoma.
5. Which states contained no tornadoes, according to this data set? According to this data set, every state experienced a tornado.
6. Create a definition query on the data and select out the January tornadoes, and examine the pattern. Repeat for the April tornadoes, and then the July tornadoes. Does the pattern support the discussion earlier in this chapter about the seasonal pattern of tornadoes? Yes, tornadoes seem to be most numerous in January and February near the Gulf Coast, and expand in number and in areal extent to the Dakotas by late summer, and ebb back down to the Gulf by the end of the year. Some exceptions exist, but the ebb and flow of the annual pattern is persistent, consistent with our discussion on air masses and fronts.
7. Clear your definition query. Generate a mean center for the tornadoes. In what state is the mean center? It is in Missouri. Create a definition

query and generate a mean center for only the F4 and higher tornadoes. Compare the locations of the two mean centers. The location of the mean center of the F4 and higher tornadoes is also in Missouri but about 90 miles northeast of the mean center for all of the tornadoes.

8. Clear the definition query. Create a standard deviational ellipse considering all the tornadoes. Approximately how many states are covered wholly or in part by this ellipse? It appears that most states, about 35, are covered at least partly by this ellipse, which extends from Virginia to Colorado. This reflects the widespread distribution of tornadoes in the United States. Repeat for the F4 and higher tornadoes, and compare the two ellipses. The standard deviational ellipse for the F4 and higher tornadoes is centered in Missouri but has a much smaller size, reflecting the Great Plains and central lowlands of the country as the focal point for intense tornadoes; this ellipse covers about 20 states from Ohio to Kansas, and from Wisconsin to Louisiana.

9. Generate a mean center and a standard deviational ellipse for January, April, July, and October's tornadoes. Compare these ellipses in terms of shape and size. Name two reasons why the ellipses vary so much during each season. The ellipses vary so much because of the air masses, driven by seasonal change and prevailing winds, that change the pattern of the tornadoes from winter to summer and back to winter again.

10. Zoom into Ann Arbor, Michigan, and select the tornadoes that have occurred in that area. Generate a set of buffers around the tornadoes in that area. Identify areas that are outside of a distance that you choose, say 1000 feet or 1000 meters, from a tornado having occurred there over the span of this data set. Do any such areas exist? Yes, depending on where you select the tornadoes, some areas will be outside the buffer distance. Then, answer the following question: Do such areas that exist outside the buffers mean that those areas are relatively safe, or does it simply mean "it's due for a big one?" Contrary to popular belief with regard to tornadoes, earthquakes, or other natural hazards, every area that is subject to the same climate or tectonic forces is equally susceptible to another event. In other words, the areas that have not experienced a tornado in the past are not "overdue" for a tornado occurring there. Mother Nature has no "tally sheet" or GIS of past touchdowns!

8.8 Related theory and practice: Access through QR codes

Theory

Persistent archive:

University of Michigan Library Deep Blue: http://deepblue.lib.umich.edu/handle/2027.42/58219

From Institute of Mathematical Geography site: http://www.imagenet.org/

Arlinghaus, S. 2003. Tornado Siren Location: Ann Arbor, Michigan. *Solstice: An Electronic Journal of Geography and Mathematics*. Volume XIV, No. 1. Ann Arbor: Institute of Mathematical Geography. http://www-personal.umich.edu/~copyrght/image/solstice/sum03/sandy/Tornado_Siren_Location,_Ann_Arbor,_Michigan.html

Arlinghaus, S. 2001. Base Maps, Buffers, and Bisectors. *Solstice: An Electronic Journal of Geography and Mathematics*. Volume XII, No. 2. Ann Arbor: Institute of Mathematical Geography. http://www-personal.umich.edu/~copyrght/image/solstice/win01/sarhaus

Practice

From Esri site: http://edcommunity.esri.com/arclessons/arclessons.cfm

Kerski, J. Distributions, Means, and Standard Deviations. Complexity level: 3. http://edcommunity.esri.com/arclessons/lesson.cfm?id=507

Kerski, J. 10 Most Populous Cities: Spatial Statistics. Complexity level: 4. http://edcommunity.esri.com/arclessons/lesson.cfm?id=411

Kerski, J. Population Drift: Mean Center Analysis. Complexity level: 4. http://edcommunity.esri.com/arclessons/lesson.cfm?id=573

Kerski, J. Population Drift: Mean Center Analysis 1790-2010. Complexity level: 4. http://edcommunity.esri.com/arclessons/lesson.cfm?id=568

Kerski, J. Population Drift: Mean Population Center Analysis. ArcGIS. Complexity level: 4. http://edcommunity.esri.com/arclessons/lesson.cfm?id=508

Kerski, J. Stormy Weather: Tornadoes, Wind, Hail. Complexity level: 4. http://edcommunity.esri.com/arclessons/lesson.cfm?id=566

Kerski, J. Investigating Historical Tornadoes Using ArcGIS. Complexity level: 4. http://edcommunity.esri.com/arclessons/lesson.cfm?id=565

8.9 Appendix of media commentary

- *Ann Arbor News.* Saturday, July 5, 2003. Front page article by Tracy Davis, "A Pair of Emergency Sirens Added to Ann Arbor System," continued inside, with photos and map.

  Excerpt from the front page: "Sandra Arlinghaus already spent time working with computer mapping and geographic information systems at work and in her position as chairwoman of the Ann Arbor Planning Commission, so when she heard the city was trying to figure out where to place new emergency alert sirens, she put her skills to work.

  The city's own GIS mappers and emergency directors had sent staff out to gauge how well sirens could be heard in various parts of town.

  But Arlinghaus took it a step further: She mapped the locations of Ann Arbor's existing 20 sirens, the approximate areas where they could be heard and how well. The resulting map showed overlaps and gaps that helped city emergency official determine where to place two new ones this week."

- July 15, 2003. Letter of Commendation, from Congressman John Dingell, United States House of Representatives, 15th District, Michigan to Sandra Arlinghaus: "I was recently informed [of] your work mapping Ann Arbor's emergency sirens. With your computer mapping project, the city will be able to determine where to build their new sirens, notifying those currently out of earshot of an emergency. Your effort is a sign of your dedication to the people of Ann Arbor and their safety. Thank you for going beyond the call of duty."

- *Ann Arbor News.* Thursday, July 19, 2003. Opinion Column by Judy McGovern, "Cheers and Jeers," on the Editorial Page, A8. "Cheers: Sandra Arlinghaus for going beyond her duties as Ann Arbor Planning Commission chairwoman by mapping on computer the location of Ann Arbor's 20 emergency sirens. Her work showed city official where the sirens could not be heard and helped them establish locations for two new sirens."

- March 11, 2004. Sandra Lach Arlinghaus, recipient, *The President's Volunteer Service Award.* President George H. W. Bush. One of a number of such awards granted to local citizens in a ceremony hosted by the City of Ann Arbor Neighborhood Watch Coordinator, Adele El Ayoubi, with certificates presented by the Ann Arbor Chief of Police, Daniel J. Oates.

Map Projections

Keywords: Projection(s) (133), map (66), globe (26), plane (24), parallels (20)

A map is the greatest of all epic poems.
Its lines and colors show the realization of great dreams.

Gilbert H. Grosvenor, Editor
National Geographic Magazine, 1899–1954

9.1 Introduction

The only commonly used, dimensionally true representation of the Earth, free of distortion, is a globe. Maps, whether on paper or on a computer screen, are flat. The transformation by which geographic locations (latitude and longitude) are sent from the curved, two-dimensional, surface of a sphere to a flat, two-dimensional, map is called "projection." The transformation preserves two dimensionality.

Imagine the Earth as an inflated balloon. Cut it open and flatten it. It will be stretched in some places and shrunk in others. Distortion is inevitable. Since the transformation from a bounded, curved two-dimensional surface to an unbounded, flat two-dimensional surface is not a smooth one, every map projection is distorted in at least three, and sometimes in four, of the following properties: Shape, area, distance, and direction. Thus, the challenge to the mapper in choosing appropriate projections, as in selecting colors, data sets, and a host of other variables, becomes, "but does it fit or is it appropriate?" Readers will have a chance to put theory into practice at the end of this chapter: Look for example at links associated with the Mollweide projection in the theory section and then practice selecting it, or other projections, in the later practice activities. Readers with even deeper interests might wish to delve into the vast literature on the topic of map projections and cartography, some of which is cited at the end of this book. Let us see why map projection selection might (or might not) matter, beginning with a real-world story with some serious implications.

9.2 In the news...

In 2009, a fascinating blog appeared on the Esri website, apparently derivative of an event associated with an Esri International User Conference (Maher, 2009: http://blogs.esri.com/esri/arcgis/2009/07/15/the-buffer-wizard-in-arcmap/). Esri ArcMap Team member, Tom Bole, had spotted a map that was published in *The Economist*, May 1, 2003, which attempted to show the countries that could be reached by North Korean missiles of different ranges. The map was made simply by buffering Pyongyang, the capital of North Korea, without regard to the underlying projection (a Mercator projection apparently had been used). The result was to suggest that all of Europe, most of Africa, and most of North America were not reachable by these missiles. Common sense might suggest that clearly all of Alaska is within 10,000 kilometers of North Korea. Indeed, when Bole checked distances, he found that the map portrayed a seriously erroneous impression. He corrected the map, using a different underlying projection—one on which buffer appearance and buffer distance aligned accurately, and sent the correction to *The Economist* and it was published soon after, on May 15 of the same year (http://www.economist.com/node/1788311). The original choice of map projection left most North American readers with the impression that they were outside the range of North Korean missiles. When the underlying projection was corrected it became clear that any feeling of confidence about being out of range was wrong. In local studies projection choice is often unimportant; in global studies, however, a bad choice can produce devastating misinterpretation (Maher, 2009). The phrase, "Don't believe everything you read" extends to maps, as well! In order to avoid making bad choices, we also consider how projection distortion might be sampled, in advance, so that the map user has an idea of what map projection to choose and what not to choose.

9.3 Looking at maps and their underlying projections

The creation and classification of map projections is a broad field, and the decision of which one to choose for any particular application or problem is a complex one. Often it is helpful to have a set of visual cues in making mental comparisons of map projections. Visualize the graticule of the globe. Compare the grid arrangement of parallels and meridians on the map with your mental image of the graticule arrangement on the globe. On the globe graticule there are a number of simple geometric properties—are these preserved on the map you are looking at? On the globe:

- Parallels are spaced equal distances apart.
  - The Equator is the longest parallel and is unique in that regard. It is the only parallel representing a great circle on the sphere.
  - Lengths of parallels, representing small circles, decrease as one moves toward the poles. The pattern of decrease is symmetric on either side of the Equator.
- All meridians are of equal length.
  - They are halves of great circles extending through the north and south geographic poles.
  - They converge at the poles.
  - At given latitude, meridians are evenly spaced.
- Parallels and meridians meet at right angles.

One commonly used method of classifying map projections groups them by the type of surface onto which the graticule is theoretically projected **(Figure 9.1)**.

- *Cylindrical.* Consider a light bulb in the center of a globe. Wrap a sheet of paper around the globe, tangent to the Equator. Meridians and parallels from the globe are projected onto the resulting cylinder as straight, parallel lines. These sets of lines meet at right angles on the cylinder, which becomes the map, just as they do on the globe. The meridians on the projected cylindrical surface do not meet at the poles, as they do on the sphere. Thus, maps made in this way are increasingly stretched and distorted toward the poles. The left-hand column of the figure **(Figure 9.1)** shows the graticule projected to a cylinder in "regular" position (with the axis of the cylinder passing through the Earth's polar axis), to a cylinder in "transverse" position (with the axis of the cylinder orthogonal to the Earth's polar axis—the basis for the UTM, or "Universal Transverse Mercator" grid), and to a cylinder in "oblique" position (with the axis of the cylinder inclined at any other angle to the Earth's polar axis).

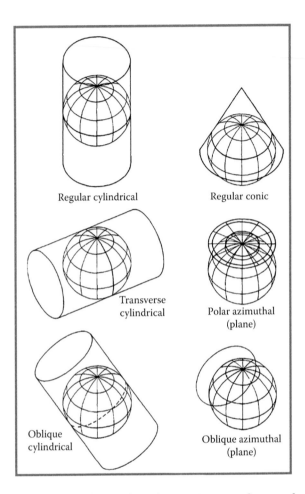

Figure 9.1 *Conceptual view of a number of projection types for transforming the graticule to a grid. Source: Snyder, J. P. 1987. Map Projections—A Working Manual, US Geological Survey Professional Paper 1395. http://pubs.usgs.gov/pp/1395/report.pdf*

- *Conic.* Imagine a cone of paper placed over a globe, tangent to the globe along one line (small circle). Often the apex of the cone is above the north geographic pole, creating a visual image with the globe of an upside-down ice cream cone. The graticule on the sphere is projected onto the cone that was rolled from a piece of paper. In the right-hand column of **Figure 9.1**, the top image shows this concept. As with the cylinder, the axis of the cone may be inclined at a variety of angles to the Earth's polar axis.

If the small circle that is common to both the surfaces of the cone and the sphere is a parallel (line of equal latitude), as it will be if and only if the conical apex lies on the extended polar diameter of the Earth, then the projection that results when the cone of projection is unrolled is called a conic projection with one standard parallel (the common small circle). If, instead, the cone slices through the Earth, so that there are two standard parallels, the resulting projection is called a conic projection with two standard parallels (**Figure 9.2a**). This projection is well-suited for showing areas in middle latitudes with predominantly east-west extent (as distortion increases north and south with increasing distance from standard parallels) such as the United States of America.

Generally, anywhere the projection surface touches (is tangent to) the globe, scale is true. This situation can occur at a point, or along one or two lines. These lines are called standard lines, and if the lines are parallels, they are called standard parallels. Distortion, or deformation of shapes, increases with increasing distance from a standard point or lines.

One might wonder why a cone and a cylinder are coupled with the natural choice of a plane, as surfaces onto which to project the Earth's graticule. The answer is simple: They are the so-called "developable" surfaces. They can be transformed to a plane without distortion. This can be done as follows: Make a cut the long way on a cylinder and unroll it into a plane; or, make a cut from the apex to the base on a cone and unroll it into a flat map (Hilbert and Cohn-Vossen, 1952). In an earlier chapter, we mentioned mapping on a Möbius strip; the reason we did so is that the Möbius strip is also a developable surface, as is the "Klein" bottle. The two latter surfaces are difficult to imagine and are seldom used in practice. They are, however, quite interesting in theory. Sometimes it is convenient to group maps by whether or not they were created on developable surfaces.

- *Azimuthal.* Now, instead of a cone or a cylinder, imagine a plane tangent to a point on the globe. The plane will serve as the projection surface. When the point of tangency is a pole, meridians projected from the spherical surface onto the plane are straight lines radiating from the pole. The last two images in the right-hand column of **Figure 9.1** show the concept of an azimuthal projection in which the graticule on the sphere is projected to a grid on a plane. Again, the relative orientation of the polar axis and the tangency point of the plane may vary. An azimuthal projection correctly shows directions (azimuths) from a single point to all other points on the map. **Figure 9.2a–d**, shows three azimuthal projections, each centered on the North Pole, as polar, oblique and equatorial views. Note the difference in the spacing of the parallels and the direct implications for the appearance of the landmasses. Learn to look at all maps by first studying the grid pattern to see how it will affect the coastlines or other boundaries.
  - If parallels are drawn in equally spaced concentric circles centered on the tangent pole, then this projection is known as an azimuthal equidistant projection. It shows the distance between any two points on a line through the center at true scale. Distances and directions of measurement to all places are true only from the center point. The polar view map in **Figure 9.2e** is centered

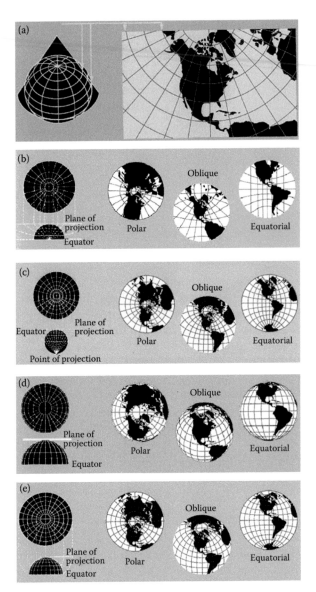

Figure 9.2 *(a) Conic projection based on two standard parallels. (b) Gnomonic projection, polar, oblique, and equatorial positions. (c) Stereographic projection, polar, oblique, and equatorial positions. (d) Orthographic projection, polar, oblique, and equatorial positions. (e) Azimuthal Equidistant projection, polar, oblique, and equatorial positions. QR code to first linked source appears later in this chapter. Source: http://egsc.usgs.gov/isb/pubs/MapProjections/projections.html. Source for extra views: Internet Archive WayBackMachine: Map Projection Pages, 2005, http:// web.archive.org/web/20070104121606/http://www.3dsoftware.com/Cartography/ USGS/MapProjections/. Images based on material from Snyder, 1987 and Snyder and Voxland, 1994. Extra material (Weisstein, 1999: http://mathworld.wolfram. com/ConicProjection.html.)*

on the North Pole. It is the map projection (when centered on Pyongyang) that should initially have been used to measure true nuclear range distances from a center point of Pyongyang!

- If, on the other hand, parallels are also projected, then the point from which they are projected will produce maps with differing spacings of parallels, different coverage of area, and of different appearance. Theoretically, there is an infinite number of points which might serve as centers for projection. These are often referred to as "perspective" projections. They are easy to visualize geometrically and thus seem particularly appropriate to consider in a book on spatial mathematics! Three common ones are:

  - *Gnomonic* (**Figure 9.2b** shows a polar gnomonic): The center of projection is the center of the sphere. The derived map covers less than the surface of a hemisphere and is bounded by, and does not include, the Equator. Great circle routes on these maps appear as straight lines; useful for planning routes of various sorts.

  - *Stereographic* (**Figure 9.2c** shows a polar stereographic): The center of projection is the pole opposite the tangent pole. The derived map covers all points on the spherical surface except the projection pole. Thus, if the projection pole is the North Pole, and the projection plane is tangent at the South Pole, all points on the spherical surface, except the North Pole, project in a one-to-one transformation to the plane. Do you now see why, in **Chapter 4**, it takes the same number of colors to color a map in the plane and on the spherical surface?

  - *Orthographic* (**Figure 9.2d** shows a polar orthographic): The center of projection is the point at infinity opposite the tangent pole. The derived map covers all points on a hemispherical surface and is bounded by, and includes, the Equator.

Map projections that do not fit within these three classes of cylindrical, conic, or azimuthal are described as Pseudo or Miscellaneous projections. Though few maps are truly the result of such simple geometric projection (most are derived from mathematical formulas), the geometric strategy is a useful way to visualize and understand the transformation process. Many maps are actually derived from mathematical formulas, which in some way imitate elements of these processes, but these calculated maps are not true geometric projections. Before moving on, let us first consider a method to formally capture what we can all see intuitively in maps of varying underlying projections: Distortion!

9.4 Sampling projection distortion

The Tissot Indicatrix is the classical way to sample projection distortion (Tissot, 1881). A sequence of circles of constant radius is centered on graticule intersection points. The greater the associated distortion in mapping the globe to the plane, the greater the distortion of the circle, which will be shown either as an enlarged circle or as an ellipse with long major axis in relation to its minor axis.

The Mercator projection is based on the geometrical idea of a cylindrical projection. **Figure 9.3** shows how a Mercator projection distorts the relative sizes of land areas on the Earth: As one moves toward the poles, areas are increasingly enlarged. The Tissot circles correspondingly enlarge with movement toward the poles. They do indeed reflect the underlying landmass enlargement associated with the manner of projecting the surface of a sphere to the plane. Choose a Mercator projection as a navigation chart at sea—it shows true compass bearings. Do not choose a Mercator projection to make land area comparison—Greenland will be larger than all of South America!

Figures 9.4 and 9.5 reveal elongation (enlargement and shape distortion) of Tissot ellipses on two different pseudocylindrical projections: Mollweide and sinusoidal, respectively. What differences do you detect in the pattern of meridians in these last two figures? Note that in the case of the Mollweide, the meridians are halves of ellipses while in the sinusoidal, they are halves of one period of a sine wave. What are their implications for mapping? Note that one of them shows greater compression at the poles than the other. At

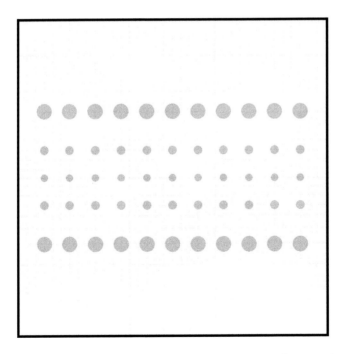

Figure 9.3 *Tissot's Indicatrix, Mercator projection. Source of original image: Eric Gaba, June 2008. Data: US NGDC World Coast Line (public domain). Extra material (Weisstein, 1999: http://mathworld.wolfram.com/MercatorProjection.html).*

Figure 9.4 *Map of the world in a Mollweide projection with Tissot's Indicatrix of deformation. Each red circle/ellipse has a radius of 500 kilometers. Scale 1:5,000,000. Source of original image: Eric Gaba, June 2008. Data: US NGDC World Coast Line (public domain). Extra material (Weisstein, 1999: http://mathworld.wolfram.com/ MollweideProjection.html).*

Figure 9.5 *Map of the world in a Sinusoidal projection with Tissot's Indicatrix of deformation. Each red circle/ellipse has a radius of 500 kilometers. Scale 1:5,000,000. Source of original image: Eric Gaba, June 2008. Data: US NGDC World Coast Line (public domain). Extra material (Weisstein, 1999: http://mathworld.wolfram.com/ SinusoidalProjection.html).*

the edges of the projection, these observations are perhaps the most evident. Tissot indicatrices may be found for any map; the *WayBackMachine* shows them for a variety of projections (2005).

Figure 9.3 illustrates quite clearly the way in which landmasses become exaggerated in area as one moves toward the poles. **Figures 9.4 and 9.5** illustrate a change in shape, rather than in area, as one moves poleward. While the visual evidence is compelling, it is difficult to see simultaneously in the mind's eye how the graticule distortion might distort a variety of globe features. One reason for this difficulty is that these models are visually static. Indeed, in Tissot models, one sees only simple country-level boundary distortion in association with ellipse elongation—at the scale of the full globe it is difficult to see more.

9.5 Some projection characteristics

With a clearer view of visualizing distortion at the global scale, we consider once again a few common projection characteristics and associated distortion issues.

- *Conformal* (**Figure 9.3**). A conformal projection maintains shape fidelity in local (small) areas. No map preserves the shape of global (large) areas. In **Figure 9.3**, the circular shape is maintained; only area is distorted. The Mercator projection is a conformal projection.
- *Equal area* (**Figure 9.4**). An equal-area projection maintains area in such a way that a small circle covering one part of the map contains the same amount of mapped area as another small circle of the same size placed elsewhere on the map. In **Figure 9.3**, Greenland appears larger than Brazil; in **Figure 9.4**, the opposite is true. It is easy to visualize the problem when the Tissot circles appear on **Figure 9.3**: Clearly, the equal-area characteristic does not apply to the Mercator projection! However, it does in the Mollweide projection where all the circles have equivalent area (but not shape). Meridians are halves of ellipses on the Mollweide. The meridians at 90 degrees east and west longitude form a circle. The Equator is a horizontal line perpendicular to a central meridian of one-half the length of the Equator. Generally, equal-area maps are good for making landmass comparisons.
- *Equidistant* (**Figure 9.5**). An equidistant projection preserves the distance from a standard point or pair of points (line). The sinusoidal projection of **Figure 9.5** is also an equal-area projection. In addition, it is an equidistant projection because distances are preserved along parallels. Meridians are halves of sine waves; as with the Mollweide, the central meridian is half as long as the Equator. No map can show the distance correctly between all points on the map.

9.6 Pseudo or miscellaneous projections

To make the image of the map look better to the casual eye, by minimizing angular and areal distortion, pseudocylindrical projections may offer a view of the Earth with evenly spaced straight parallels and curved meridians (usually equally spaced). The Mollweide (**Figure 9.4**), sinusoidal (**Figure 9.5**), and Robinson projections (**Figure 9.6**) serve as examples. Generally, bending the

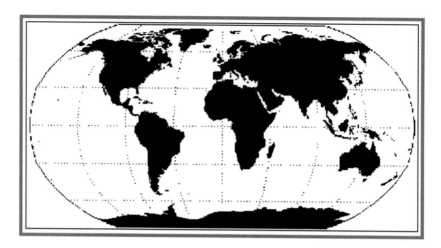

Figure 9.6 *Robinson Projection. Compromise projection adopted by the National Geographic Society to serve as a base map for many of their world maps, over a bounded time period. Found originally on the link listed below. NASA, Pseudocylindrical Projections. http://idlastro.gsfc.nasa.gov/idl_html_help/Pseudocylindrical_Projections. html*

graticule as one proceeds poleward carries the map "flesh" on the "bones" of the graticule to a more globe-like appearance.

The idea of using an underlying grid system to transform one set of masses to another is not unique to geography or mapping. In the early part of the twentieth century, Sir D'Arcy Wentworth Thompson (1917) employed a similar strategy to look for mathematical characterizations that would carry one fish species to another. The cartographer Waldo Tobler (1961a and 1961b) carried this idea forward in the world of mapping. Indeed, Tobler's hyperelliptical is another compromise projection (pseudocylindrical) that has the additional characteristic of being equal-area (http://www.csiss.org/map-projections/Pseudocylindrical/Hyperelliptical.pdf).

There are an infinite number of maps possible and thus an infinite number of categorization schemes possible. There is no "best" set of categories. Some might be more useful than others, though. Another broad type of projection that the reader may see is "interrupted" projections. These seek to emphasize landmasses by cutting the projection apart in the ocean. The Goode homolosine, based on the sinusoidal and the Mollweide projections, is one example, shown in **Figure 9.7**. The use of the Mollweide in latitudes closer to the poles reduces the meridian compression evident in the sinusoidal. Others include the Philbrick interrupted sinumollweide, also based on the sinusoidal and the Mollweide projections (Philbrick, 1986 reprint). Abstractly, all maps are "interrupted"—but when the interruption occurs along rectangular outer edges, our rectilinearly trained minds are not jarred!

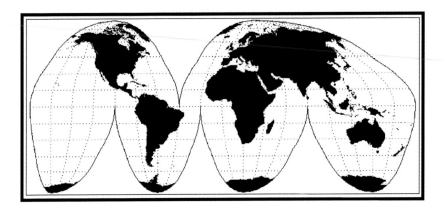

Figure 9.7 *Goode's Homolosine Projection. Source: NASA, Pseudocylindrical Projections. Found originally on the link listed below. http://idlastro.gsfc.nasa.gov/idl_html_help/ Pseudocylindrical_Projections.html*

9.7 Contemporary approach: Web Mercator Auxiliary Sphere projection

Many contemporary mapping packages employ a simple grid transfer from a sphere to a plane to portray the graticule and the attached landmasses. Naturally, distortion is severe; what is valuable is that this particular characterization is easily projected into existing projections. When one begins with a base, such as the Robinson projection, that ease of reprojection is lost as numerical truncation of infinitely repeating decimals may cause a "rupture" in the reprojection process forcing a crash of hardware.

As times change, so do different needs for projections. In the contemporary world of online mapping, the Web Mercator Auxiliary Sphere projection has become the default projection of Google Maps, Google Earth, ArcGIS Online, and no doubt other related services already online or yet to come. This projection is a Mercator-like projection that treats the two-dimensional curved surface being projected as that of an oblate spheroid rather than as that of a sphere. Its advantage is straightforward—most major online mapping tools use this projection. Therefore, thematic data from one mapping platform, such as world ecoregions or ocean currents, can be easily moved to another mapping platform. Its disadvantages are also straightforward; it is a Mercator projection and has the same limitations.

Alistair Aitchison (2011) offered an interesting example, in terms of maps of Great Britain, of the differences that come from using a web Mercator projection based on a spherical surface and one using an oblate spheroidal surface. The differences are small, but discernible. When the sphere's polar axis is squashed, the landmasses on the surface are pushed more toward the Equator than when it is not. Issues involving large scale mapping and projections may play out in related, but not identical, ways in association with shape choice of the underlying projection surface—the differences may or may not be discernible.

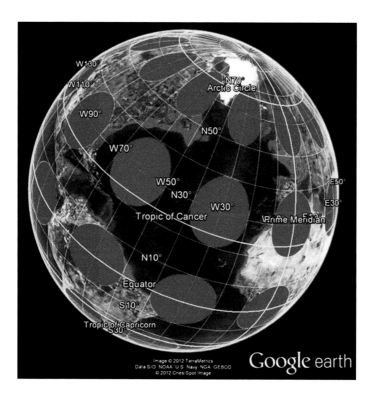

Figure 9.8 *Tissot Indicatrix of Web Mercator Auxiliary Sphere projection with distortion realized in Google Earth. Source: Based on Google Earth mapping service © 2012 Google and Image © 2012 TerraMetrics, Data SIO NOAA US Navy NGA GEBCO, ©2012 Cnes/Spot Image. Altered base image from Arlinghaus, S. L. and Kerski, J. 2012.*

Software in which one can traverse an image of the globe, such as Google Earth, offers a dynamic way to both see Tissot-style distortion and simultaneously consider an inventory of what is available on the globe through digital imagery. **Figure 9.8** shows a visual sampling of projection distortion, using Tissot's Indicatrix, of the Web Mercator Auxiliary Sphere projection (Arlinghaus and Kerski, 2012). What appears to be poleward distortion in the screen capture in the figure will change as one rotates the linked image within the Google Earth software. Why is that? The figure in the book is a static image. Try moving (using a link at the end of this chapter) the Google Earth surface around to see what happens to the pattern of red circles!

9.8 Sampling the environment: The degree confluence project

The Tissot Indicatrix samples the graticule at the global, or at least, the small-scale level. A more recent project, called the Degree Confluence Project (http://www.confluence.org), photographically samples the Earth's graticule at integer locations of parallels and meridians to create an archived inventory of plants, animals, landforms, land use, and human impact at these locations. Consider individual photos and reports of the set of discrete locations (Kerski confluence visit at 42 degrees north latitude, 84 degrees west longitude, a sample typical of such visits; Kerski, 2010). Each one gives an amount of geographic information confined by latitude and longitude within a small capsule. When one is standing on the spot marking a degree confluence, projection does not matter. Recall our earlier discussion on datum: Datum certainly matters because the datum used affects the position read by the GPS receiver. Therefore, the position marking the intersection of the latitude and longitude lines could be in a different meadow hundreds of meters away depending on the datum used. But projection does not matter at this local scale. When one wishes, however, to look at the whole set of confluences at a smaller scale, projection does matter. As the Tissot ellipses mesh, as one moves poleward in the sinusoidal projection of **Figure 9.5**, so too do the integer latitude/longitude intersection points and associated photographic inventory. Converging meridians, as one moves poleward, force this action.

In previous chapters, we have paid little attention to projection. For local studies, projection considerations are not as important as they are for global studies. As we function in the world, we are not generally aware of the curvature of the Earth. A good general rule seems to be to choose the projection that best fits your needs: If comparing landmass areas across the globe, for example, choose an equal area projection. In global studies covering a broad territorial expanse, projection selection can make a huge difference in the portrayed outcomes.

9.9 Practice using selected concepts from this chapter

9.9.1 Overview

The material in this section presents an opportunity to view different maps, cast in different projections, using dynamic maps in a web browser. This process is done via live GIS mapping services in ArcGIS Online, a cloud-based GIS used elsewhere in this book. Here, the reader can use this platform to map the location of global natural hazards, such as earthquakes, volcanoes, and tsunamis, on different projections. By varying the projection, not only do the differences in land and ocean patterns become apparent but so do the interpretations of those patterns in relation to natural hazards.

9.9.2 Comparing projected data using ArcGIS online

To participate in this activity, go to ArcGIS Online (http://www.arcgis.com/home). Click on the Groups tab and, in the search slot, type in the group name "Projected Basemaps" (without the quotation marks...note that 'basemaps' is a single word). In the search results window, click on the thumbnail image to open the group. Scroll down and open the map "Goode Homolosine Land" in the ArcGIS.com map viewer. The Goode homolosine projection is a pseudocylindrical, equal-area, composite map projection used for world maps. Its equal-area property makes it useful for presenting the spatial distribution of phenomena. It is typically presented with multiple interruptions, as is shown in this map, with the "interruptions" being the oceans. Click on "Add to this Map." In the "Find" slot, search for data "Global Natural Hazards Data," and in the results window, Add the Global Natural Hazards Data by NOAA_NGDC (marine geological samples). Select "Done Adding Layers." Expand the layer and make the data visible as "Significant Earthquakes." Click on the map legend button to view the legend. The western hemisphere of a map generated using this process appears in **Figure 9.9** (please be aware that web-based mapping may change; interpret the specific directions as needed in order to generate a map similar to this one).

Describe the pattern of significant earthquakes around the world. Look particularly at coastal areas. Do you find regions that have had larger earthquakes that are a surprise to you? Consider the large orange circle in the interior of the USA, away from the coast (**Figure 9.9**). That situation may surprise a number of North American residents. It illustrates, however, the

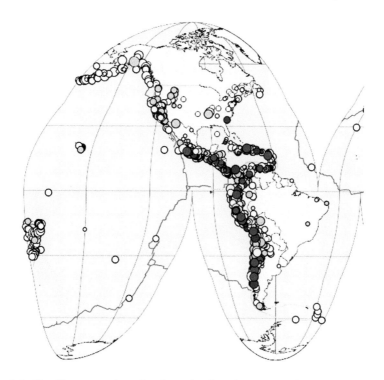

Figure 9.9 *Significant earthquake data displayed on a Goode's Homolosine projection (cropped to show the Western Hemisphere). Source: Esri base map.*

power of using a map as a guide to asking research questions. Maps can serve as far more than mere displays to support text. When used well, they can also guide research.

Consider the global pattern of earthquake data when displayed on a Goode's homolosine (**Figure 9.10a**). Note that there are numerous events around the margin of the Pacific Ocean. Yet, because the projection is interrupted in the oceans, any unified pattern surrounding that ocean is difficult to visualize. Go back and open the map "coordsys_EckertIV_150E," the Eckert IV projection (scroll down or go to extra pages as needed). The Eckert IV projection is a pseudocylindrical map projection, in which the length of Prime Meridian is half that of the Equator, and the lines of longitude are semiellipses, or portions of ellipses. As you observe this map, consider the significance of the "150 E" in the title of this map file. Since this map is centered on a meridian

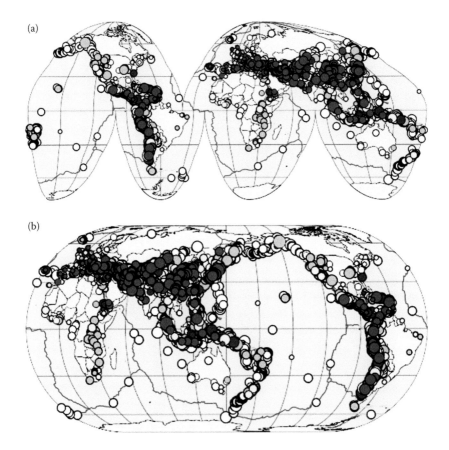

(a)

(b)

Figure 9.10 *(a) Global view of earthquake data displayed on a Goode Homolosine projection. (b) Global view of earthquake data displayed on an Eckert IV projection (centered on 150E). Source of base map: Esri software.*

that is centered in the Pacific Ocean, the Pacific basin now becomes the prominent feature on the map. As before, add the global natural hazards data from NOAA/NGDC (NOAA/National Geophysical Data Center) and make the significant earthquakes visible (**Figure 9.10b**). Now, the earthquake pattern around the Pacific rim comes to the fore. The homolosine was good for looking at the interior of landmasses but was not good for looking at patterns involving the land/ocean interface. Since the Eckert IV could be easily centered on the 150 E meridian, and did not interrupt the Pacific Ocean, it was vastly superior to the homolosine for visualizing earthquake pattern in relation to the Pacific Ocean.

Go back to the group and open the map "Bonne_Projection" (a sample appears on the title page of this book). This map may be on the second page of the Projected Base maps group. A Bonne projection is a pseudoconical equal-area map projection that is memorable for its "heart" shape. The map may come up showing only the oceans (**Figure 9.11a**). As you did before, add the global natural hazards data and make the significant earthquakes visible; but first, add a layer showing the landmasses (**Figure 9.11b**). How does this projection help you understand the global distribution and pattern of earthquakes? Consider the location of earthquakes in relation to the tectonic plate boundaries, clearly visible in **Figure 9.11a**. The oceans are important, but aren't other features just as important? Certainly they are in the case of studying the relationship of the tectonic plate boundaries to earthquake location. In terms of broad general context, however, the land/water interface offers enhanced visualization!

As we have compared and contrasted several projections and consider their merits and drawbacks for visualizing and understanding the earthquake data at the global scale, so too we hope the reader will take this sort of critical spatial analysis into other spatial realms.

9.10 Around the theoretical corner?

By now it should be clear from the discussions in this book that there is no perfect map. This is also one of the most interesting things about maps. Maps may not be perfect, but they are incredibly useful. The myriad possible map projections, each useful to somebody or some application, is a perfect example of the utility of maps. There are thousands of different map projections and an infinite number more yet to be discovered. By the end of 2012, for example, ArcGIS for Desktop software supported 4634 different map projections. The number of map projections might appear a daunting prospect; it is, however, a stimulating prospect! There is always something new around the corner, yet to be discovered, in the wide world of mapping. Why, one might ask, are there an infinite number of possible projections available?

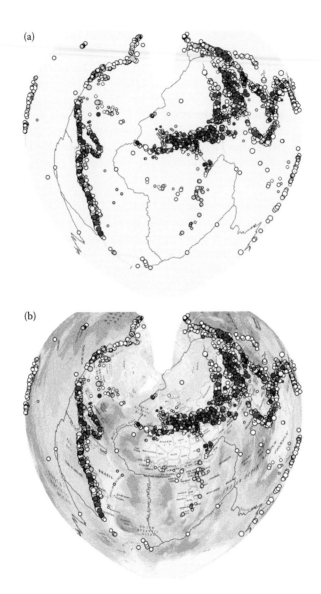

Figure 9.11 *Bonne projection, (a) oceans and tectonic plate edges only, and (b) oceans, tectonic plate edges, and landmasses. Source of base map: Esri software.*

The answer lies in the fact that the geometry of mapping is assumed to be Euclidean.

The goal of mapping the surface of the Earth-sphere (globe) to the plane is to do so in a manner that is one-to-one: No point on the globe corresponds to more than one point in the plane, and vice versa. Stereographic projection of the spherical surface (from the North Pole) to a tangent plane (tangent at the

South Pole) projects the entire spherical surface, except one point—the North Pole—to the plane. The reason that the North Pole does not project into the plane is that the line projecting the North Pole runs parallel to the tangent plane. In Euclidean geometry, a line parallel to a given plane never intersects it. Stereographic projection is the best we can do, in terms of getting all the points on the globe (but one) into the plane. While this feature of the stereographic projection might seem to make it attractive, the distortion introduced in this projection might not be desired in a number of applications.

Euclid's Parallel Postulate states that given a line *l* and a point *P* not on *l*, there exists exactly one line *m* through *P* that does not intersect *l*. It is more common perhaps to think of this as "parallel lines never meet." Euclid's Parallel Postulate is behind almost all of the decisions we make in contemporary mapping and is responsible for the infinity of selections available. It has been with us for thousands of years and our minds are conditioned through intellectual institutions to think naturally in a world that embraces it as a foundation.

There is, however, a whole class of geometries that deliberately violate this postulate (Coxeter, 1965; 1998). If an infinite set of points at infinity is added to the plane, then the line through North Pole (the "parallel") and a line in the tangent plane intersect at a point at infinity. The concept of "parallel" has been discarded. This sort of geometry, based on a simple violation of Euclid's Parallel Postulate, which denies the presence of parallels, is called "elliptic" geometry. So, we have a situation with no parallels (elliptic geometry) and with a unique parallel (Euclidean geometry). We can also have a situation with multiple parallels. Imagine multiple lines passing through *P* that bend and become asymptotic to *l* (**Figure 9.12**).

These lines are all "parallels." This sort of geometry, based on violating Euclid's Parallel Postulate from above (more than one parallel to a given line) is called "hyperbolic" geometry. These local geometries have found application in physics and some other contexts (Minkowski, 1907/1915). What might be discovered by such alteration of the foundations of the world of mapping?

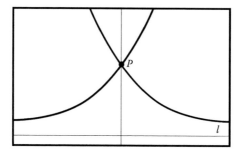

Figure 9.12 *Two parallels to line l pass through point P. Based on material from Arlinghaus, S. L. 1986. "The Well-tempered Map Projection." Essays on Mathematical Geography. Monograph #3, pp. 1–27. Ann Arbor: Institute of Mathematical Geography. http://www-personal.umich.edu/%7Ecopyrght/image/monog03/fulltext.pdf*

9.11 Exercises

9.11.1 Overview

This section outlines two exercises in detail and suggests a few hints for solution to get the reader started on the right track. To see the detailed solutions, consult the associated Solutions Manual. These exercises are designed with the reader in mind who has access to high-end software and computing equipment. Following them, a set of links only to some existing online exercises offers further challenges to the reader.

9.11.2 Comparing projections with ArcGIS for desktop

In previous material in this chapter, you used a web-based GIS to compare data sets mapped onto three different projections. These map projections were already set up, and you used them to analyze how different data appears mapped on these projections. In this exercise, you have the opportunity to change the map projections. As this is an even more powerful technique, you need a more powerful set of tools to accomplish it, which are found within ArcGIS for Desktop software. You will examine the shape of countries, plate tectonics, and the world graticule in unprojected space, in the Eckert IV projection, in the Robinson projection, and in the Mollweide projection.

Go to ArcGIS Online (http://www.arcgis.com/home) and, after making sure that your search criteria are set to "all content" instead of "web content only," search for the following data: "World Earthquakes Volcanoes Faults Countries Plate Polys Elevation owner:jjkerski." This is a map package, which contains all the data you need to explore projections in a rigorous way. Under the thumbnail image, select "open in ArcGIS for Desktop." Once the map opens, click the globe tool to zoom to the extent of the map. Turn off the elevation image to get a clearer view of the countries. Select "View→ Data Frame Properties" and then "Coordinate System" to check the map's current projection. Note that it is listed as "geographic, world, WGS 84." The map is not in a projected coordinate space; it is a latitude–longitude representation of data in a "geographic" coordinate space. Examine the shape of the countries on the map. What is a disadvantage of this geographic coordinate space view in terms of the shape of countries near the North Pole and the South Pole? In light of what you have learned in this book, what are some of the dangers inherent in using data in this geographic coordinate space? The issue of spreading or compression of the graticule near the poles is a significant one that can impact the visualization of data.

Use the Add button to Add Data from ArcGIS Online. Search ArcGIS Online for "Grid" and add the "World Latitude and Longitude Grids" layer. This is the 1 degree grid for the world. Describe the latitude and longitude lines as they appear on this geographic projection. At what angles do they intersect? Use the measure tool to measure the distance between two specific lines of longitude.

Use the Data Frame Properties to change the map projection. Change it to Projected→ World→ Eckert IV and consider the same issues as in the last paragraph. Repeat the process for the Robinson projection. Consider the variation among these three projections.

Change the projection to Mollweide.

Do some research on the Mollweide map projection at sites such as the following linked site embedded also in the QR code: http://www.diversophy.com/petersmap.htm; or, at the next link, also embedded in the next QR code

http://www.research.ibm.com/dx/proceedings/cart/cart.htm. On the Mollweide projection recall from above that meridians are formed from ellipses.

Review the USGS map projections poster (linked below and also through the QR code), and observe how Mollweide and Robinson fit into the categories listed on the poster: http://egsc.usgs.gov/isb/pubs/MapProjections/projections.html

Change the projection to Projected→ World—The World from Space. Note that this is an orthographic projection. This perspective projection views the globe from an infinite distance, giving the illusion of a globe. This projection is therefore often referred to as "the world from space" because the globe in this projection has this appearance. Select Apply but not OK yet to keep the coordinate system box visible along with the map. Right click on the projection name and select "copy and modify" and customize the orthographic as follows: Change the Center Longitude to –75, the longitude of the eastern USA. Select OK and look at your map. Think about what center latitude and center longitude the projection should be set to in order to center the map on Egypt.

9.11.3 Internet café in Denver activity

In this exercise, you will have another opportunity to do some hands-on work with map projections, using data that you worked with previously to locate the optimal site for an Internet café in Denver. Creating areas of proximity, or buffer zones, around points, lines, or areas is dependent on the underlying map projection of those points, lines, or areas. Since the buffer tools in GIS software use distances and areas to calculate and draw the zones, the shape and size of the zones are affected by the map projection used. In the hands-on activity below, the shape and size of the areas near schools and universities differ depending on the map projection used, and hence, the parcels of land under consideration for the café will differ depending on the projection used. A further item of interest is discoverable from the use of GIS. GIS software will calculate buffer zones even when no map projection is defined, based simply on distances in an x–y coordinate space. But as we have emphasized repeatedly in this book, great care must be used in the interpretation and use of all mapped data. Unlike the North Korean nuclear risk analysis example at the beginning of this chapter, here, buffer zones based on unprojected data may only be slightly different from those based on projected data when focusing on individual neighborhoods in a city. But even at a local scale, the inclusion of certain parcels in or out of the buffer zone takes time and money to consider, bid on, and site a business.

Recall that in **Chapter 3**, when setting up the data for the Internet café activity, you were asked to define the projection. This was done so that buffer dimensions, that should be the same size independent of their location on the map, would not vary depending on where they were placed. The results were elliptical (non-circular buffers). Go back to those directions in **Chapter 3** and repeat the exercise, this time failing to set the projection. Now, the buffers should be circular. Try setting the projection to something other than either of the two above. The result, for this "local" study, may show little difference from the WGS 84 that resulted in elliptical buffers in **Chapter 3**. Projection matters—the more global the study, the more it matters. Think about the situation that arose in the example at the beginning of this chapter, with the global map of nuclear reach centered on North Korea!

9.12 Related theory and practice: Access through QR codes

Theory

Persistent archive:

University of Michigan Library Deep Blue: http://deepblue.lib.umich.edu/handle/2027.42/58219

From Institute of Mathematical Geography site: http://www.imagenet.org/

Arlinghaus, S. L. and J. J. Kerski. 2012. From Tissot to Google Earth: Sampling the Earth's Graticule. *Solstice: An Electronic Journal of Geography and Mathematics.* Volume XXIII, No. 1. Ann Arbor: Institute of Mathematical Geography. http://www.mylovedone.com/image/solstice/sum12/Tissot.html

Arlinghaus, S. and M. Batty. 2010. Zipf's Hyperboloid — Revisited: Compression and Navigation — Canonical Form. *Solstice: An Electronic Journal of Geography and Mathematics.* Volume XXI, No. 1. Ann Arbor: Institute of Mathematical Geography. http://www.mylovedone.com/image/solstice/sum10/ZipfRevisited.html

Arlinghaus, S. L. 2007. Geometry/Geography — Visual Unity. *Solstice: An Electronic Journal of Geography and Mathematics.* Volume XVIII, No. 2. Ann Arbor: Institute of Mathematical Geography. http://www-personal.umich.edu/~copyrght/image/solstice/win07/hyperbolicgeometry.html

Arlinghaus, S. 1986. "The Well-tempered Map Projection." Essays on Mathematical Geography. Monograph #3, pp. 1–27. Ann Arbor: Institute of Mathematical Geography. http://www-personal.umich.edu/%7Ecopyrght/image/monog03/fulltext.pdf

Practice

From Esri site: http://edcommunity.esri.com/arclessons/arclessons.cfm

Kerski, J. The Geometry of Map Projections. Complexity level: 3. http://edcommunity.esri.com/arclessons/lesson.cfm?id=514

Kerski, J. Exploring Measurement — ArcGIS version. Complexity level: 3. http://edcommunity.esri.com/arclessons/lesson.cfm?id=521

Kerski, J. Exploring Measurement. Complexity level: 3. http://edcommunity.esri.com/arclessons/lesson.cfm?id=501

Kerski, J. Map Projections and GIS. Complexity level: 3, 4. http://edcommunity.esri.com/arclessons/lesson.cfm?id=593

Kerski, J. Map Projections: Why They Matter. Complexity level: 4. http://edcommunity.esri.com/arclessons/lesson.cfm?id=592

Kerski, J. Map Projections: The 1-Page Guide. Complexity level: 4. http://edcommunity.esri.com/arclessons/lesson.cfm?id=431

Integrating Past, Present, and Future Approaches

Keywords: Map (60), projection (35), point (28), population (27), route (27)

"The time has come," the Walrus said,
"To talk of many things:
Of shoes–and ships–and sealing-wax–
Of cabbages–and kings–
And why the sea is boiling hot–
And whether pigs have wings."

Lewis Carroll

10.1 Introduction

One exciting future direction for mapping, involving alteration of the heart of Euclidean geometry and consequent implications in the world of mapping, was hinted at in the last part of **Chapter 9** and earlier in this work. It is, indeed, a consideration perhaps of "whether pigs have wings." Many approaches to the "new" in fact have roots deeply embedded in earlier times. As interests and technology change, it becomes all the more important to remember not only to look to the past as an archive, but also to embrace the past and use it as a base for the future. For now, as a way to wrap up the many concepts and activities covered in this work, we offer deeper views and state-of-the-art views of the "cabbages and kings" of the contemporary mapping world and their associated mathematical underpinnings.

10.2 From the classics to the modern: Past and present

Most students of physical geography, cartography, and other subjects, including jig-saw puzzlers, become excited when they first encounter Erwin Raisz's astounding hand-drawn map of the *Landforms of the United States*. This map has served as an inspiration for generations of mappers to take the trouble to include accuracy, science, and beauty into their cartographic works. Here is a map that truly merges science and art and clearly reveals cartography as a discipline that includes both (Erwin Raisz's *Landforms of the United States*, 6th revision, 1957, http://www.raiszmaps.com/).

Evidence of this inspiration is apparent in the Thelin and Pike map made over 30 years later that shows far greater detail than even Raisz might have imagined (**Figure 10.1**). The technological revolution that is upon us permits not only clear visual improvement and archivability, but also offers user interactivity and product delivery over the Internet, as well as the ability to attach extra information on how the map was made, geological issues dealt with, and more. Today, one can download the map directly from the USGS in a variety of file formats and also follow the links to the fascinating story behind the map. **Figure 10.2** shows a screenshot of a portion of the map in its color version. Note that the map was published at a small scale, but used large scale Digital Elevation Models to create it. The resulting map shows not only broad features and processes, but also landforms only a few kilometers wide, such as Sutter Buttes in the Central Valley of California. Take the time to examine the USGS website and follow the saga of the making of this amazing 1991 edition of the *Landforms of the Conterminous United States*.

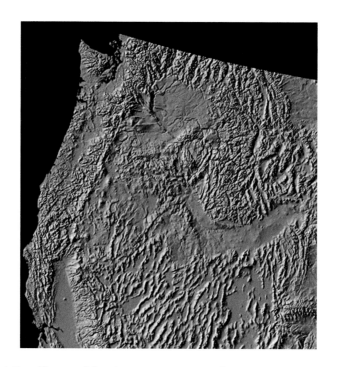

Figure 10.1 *Landforms of the Conterminous United States—A Digital Shaded-Relief Portrayal, Gail P. Thelin and Richard J. Pike. 1991. Source: USGS. http://pubs.usgs. gov/imap/i2206/*

Figure 10.2 *Thelin and Pike's Landforms of the Conterminous United States, in color USGS. Link to the full site source, http://pubs.usgs.gov/imap/i2206/.*

Yet another stunning example of converting the old to the new, and bringing out even more than anyone might have anticipated in times past, is built on ideas from physics, involving population potentials. In the 1940s, social scientists interpreted some physical principles in a variety of social contexts. In 1947, J. Q. Stewart used population potentials to depict the accessibility of people from a given point on a map. Such accessibility is a measurement of how near people are to a point; the population potential at one place is the sum of the ratios of the population at all other points to the distances from the place in question, to those points. Demographic gravitation is the physical model suggesting that concentrated groupings of people (urban populations) serve as an attractive force to pull in more people.

The 1940s work of J. Q. Stewart, as well as G. K. Zipf (1946, 1949) and others, find interesting correspondences, at least some of which appear to stand the test of time. While J. Q. Stewart interpreted gravity in terms of demography, W. J. Reilly had earlier related it to the economics of retail activity (1929, 1931). Zipf associated the concept of gravity with a dynamic "demographic energy" which led him to postulate some concepts about trip distribution through gravity models. Warntz, in 1965 created a map of the 1960 *Potentials of Population* (**Figure 10.3**). In more contemporary times, Michael Batty, University College

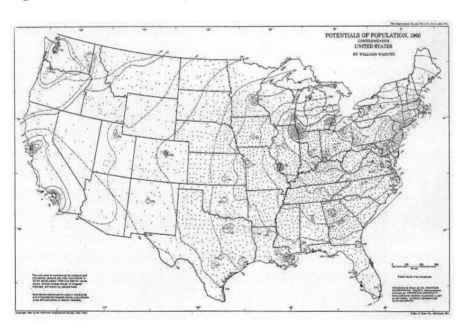

Figure 10.3 *Warntz's hand-drawn population potential map based on 1960 data (William Warntz, 1965, "A note on surfaces and paths and applications to geographical problems"). Reprinted here with permission of John D. Nystuen, Editor, Michigan Inter-University Community of Mathematical Geography, Papers, in which this map originally appeared. http://www-personal.umich.edu/~copyrght/image/micmog.html*

Figure 10.4 Grayson's interactive population potential map based on 2010 data. Source of base map: Esri software.

London, has found a number of parallels to explore between the earlier work of G. K. Zipf and spatial science (numerous works authored or co-authored by Batty). In the present, Grayson's *Potential Surface* (2012) mirrors the earlier hand-drawn maps of Warntz (1965), and of Stewarts's 1947 work (**Figure 10.4**). The underlying patterns are quite similar although Grayson's contemporary work is based on a finer mesh of pixels, uses color, and (most important) is fully interactive. The user can see different views of the population surface (**Figures 10.5 and 10.6**)!

When the scale of the Grayson map is made more local (**Figure 10.5**), it becomes easier to see the parallels with Warntz's map. Zooming in, one begins to see that what appears to be a continuous surface in **Figure 10.4** actually appears to be made up of discrete units, each of which represents 1000 persons per kilometer (with warm colors representing higher population potential). Enlarging the scale even more (**Figure 10.6**) shows a dotted surface quite similar to that of Warntz in concept although much more detailed in terms of execution.

As Grayson offered a contemporary view of J. Q. Stewart's early work, we offer readers an added opportunity to reflect on the work of Reilly, through a trucking exercise, and the work of Zipf, through a touring exercise. The thoughtful reader will no doubt look for ways to integrate these activities, laterally in geographic space, through a systematic view of network analysis and vertically, through a systematic view of map overlays and concepts involving mapping already presented elsewhere in this work.

Another innovative approach to generate and visualize geographic data is through Columbia University's Center for International Earth Science

Figure 10.5 *Screenshot of another view of the Grayson map. Source of base map: Esri software.*

Figure 10.6 *Screenshot of another view of the Grayson map. Source of base map: Esri software.*

Information Network (CIESIN)'s Gridded Population of the World (GPW) (http://sedac.ciesin.columbia.edu/gpw/). Originally produced by Waldo Tobler, Uwe Deichmann, Jan Gottsegen, and Kelley Maloy from the University of California Santa Barbara, with partial support from CIESIN under a US National Aeronautics and Space Administration (NASA) grant, the GPW is a raster representation of the global population. The purpose of GPW is to provide a spatially disaggregated population layer that is compatible with the data sets from social, economic, and Earth science fields. One of its innovations is the generation of a uniform set of population data around the world, including vast areas not covered by any national census. It accomplished this through algorithms that used input units at the national and administrative unit level of varying resolutions. With the addition of a Global Rural-Urban Mapping Project (GRUMP), the estimates incorporated the use of satellite data, such as the NASA night-time lights data set, as well as buffered settlement centroids, the technique of which should by now be familiar to the reader of this book.

The native grid cell resolution is 2.5 arc-minutes, corresponding to about 5 kilometers at the Equator, although aggregates at coarser resolutions are also provided. Separate grids are available for population count and density per grid cell. The GRUMP data provide resolution to 30 arc-seconds, which is about 1 kilometer at the Equator. The GRUMP data also provide a point data set of all urban areas with populations of greater than 1000 persons, in Excel, Comma Separated Values (CSV), and shapefile formats. In the activity section of this chapter, you will have the opportunity to examine these data sets for yourself.

10.3 A non-Euclidean future?

This book has dealt with mapping in the Euclidean world; what might happen in a non-Euclidean view of it all is a fascinating future prospect. The material below offers a non-Euclidean cartographic connection, enabling us to see all perspective projections within the framework of projective geometry. Recall that in **Chapter 9** we discussed the role of Euclid's parallel postulate in relation to the development of a class of non-Euclidean geometries. Here we probe those connections further to display a theorem involving map projection in the non-Euclidean world.

10.3.1 Projective geometry

The overarching non-Euclidean geometry, with points at infinity treated as ordinary points, is called "projective" geometry. The language of projective geometry is dual: "Two points determine a line" and its dual, "two lines determine a point." The words "line" and "point" may be interchanged in any statement that is true in the projective plane and an equally true statement will be the result. There are other pairs of dual terms, as well: "Concurrent" and "collinear," for example. The Principle of Duality is a "meta" concept about the language in which the mathematics is written. Projective geometry is a symmetric geometry that does not distinguish the ordinary from the infinite. What might be the implications of this broad geometry for the geo/metry/graphy of mapping?

10.3.2 Perspective projections

The material below shows that the entire set of perspective map projections may be derived in the projective plane. It is done using only the subset of projections with centers of projection contained within the sphere of projection. Because the transformation takes place in non-Euclidean, rather than Euclidean geometry, the unbounded problem of looking at an infinite number of projection centers spread along an unbounded ray is converted to one of looking at an infinity of projection centers spread along a bounded line segment. An unbounded, intractable problem becomes bounded, and perhaps manageable. The language of duality then applies to the geometry of all perspective map projections and offers the potential to analyze geographical problems that exhibit symmetry in underlying relations and simultaneously embrace the concept of infinity. Reduction of the complex to the simple, or the unbounded to the bounded, is a powerful ally.

In the last chapter, we displayed three azimuthal projections, in which the parallels were projected. They were gnomonic, stereographic, and orthographic projections. All three are simultaneously visualized in **Figure 10.7**. Clearly, there are an infinite number of choices available, up and down the white vertical ray, for centers of projection. **Figure 10.7** shows the relationships among the three centers of projection:

- The three projected points are collinear.
- The closer the center of projection is to the point of tangency, the farther the projected image of a point P is from the point of tangency.

It is not hard to imagine the pattern continuing in a natural way with varying choices for the center of projection on the white ray yielding expected positions on the line of projection for projections of P. A term for this style of projection, rooted in geometric language, is "perspective" projection.

10.3.3 Harmonic conjugates

One concept from projective geometry, that permits collapsing the unbounded infinity of perspective projection centers into a bounded infinity of them, involves the construction of "harmonic conjugates." The construction involves perspective projection through a point P not on a given line through points A, B, and C. It shows how to determine, uniquely, a fourth point C' that is

Figure 10.7 *Simultaneous display of gnomonic, stereographic, and orthographic projections. Source: Arlinghaus, S. L. 2007. Geometery/Geography—Visual Unity. Solstice: An Electronic Journal of Geography and Mathematics. Volume XVIII, No. 2. Ann Arbor: Institute of Mathematical Geography. http://www-personal.umich. edu/%7Ecopyrght/image/solstice/win07/hyperbolicgeometry.html*

independent of arbitrary choices made during the process of construction. The given point *C* and the constructed conjugate point *C'* are said to be harmonic conjugates with respect to *A* and *B*. The determination of the conjugate point is unique. It is independent of support points generated within the construction. The mechanics of this construction are suggested in **Figure 10.8** and the detail is available elsewhere (Coxeter, 1965). **Figure 10.9** illustrates how stereographic projection fits with the construction in the tangent plane.

10.3.4 Harmonic map projection theorem

- Centers of projection that are inverses in relation to the poles of a sphere are harmonic conjugates in the projection plane in relation to the projected images of the poles of the sphere.
- As a special case of the observation above, it follows that gnomonic and orthographic projections, with inverse centers of projection in the sphere, are composed of points that are harmonic conjugates of each other in the plane (Arlinghaus, 1986).

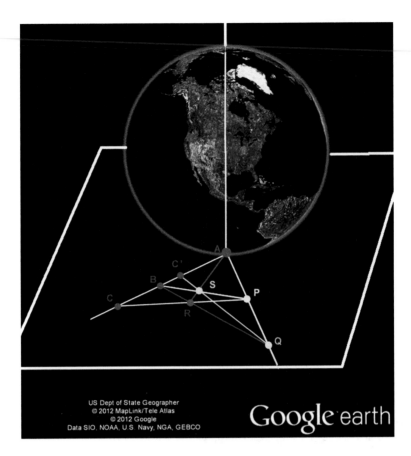

US Dept of State Geographer
© 2012 MapLink/Tele Atlas
© 2012 Google
Data SIO, NOAA, U.S. Navy, NGA, GEBCO

Google earth

Figure 10.8 On a line with three given points, A, B, and C, where A is the point of tangency of the pole with the tangent plane, construct C', the harmonic conjugate of C, as in this figure. Source: Arlinghaus, S. L. 2007. Geometery/Geography— Visual Unity. Solstice: An Electronic Journal of Geography and Mathematics. Volume XVIII, No. 2. Ann Arbor: Institute of Mathematical Geography. http:// www-personal.umich.edu/%7Ecopyrght/image/solstice/win07/hyperbolicgeometry. html

The unfolded animation sequence in **Figures 10.7 through 10.9** offers a suggested direction for the visualization of concepts. **Figure 10.10** offers some direction for proof.

The Harmonic Map Projection Theorem shows that, in the projective plane, points on a gnomonic projection are harmonic conjugates of points on an orthographic projection, with respect to the tangent pole and the corresponding point in a stereographic projection (**Figure 10.10**). Whatever we know to be true about points in a gnomonic projection is therefore known about the conjugate points in the orthographic projection. The same holds

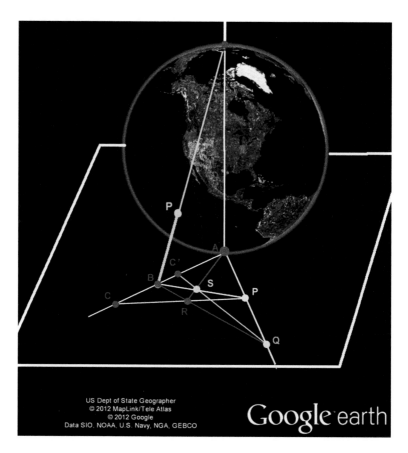

Figure 10.9 *Join B to the opposing pole (stereographic center of projection) and label the point of intersection of the sphere and the line as the point P. The point P projects to B under the transformation of stereographic projection. Source: Arlinghaus, S. L. 2007. Geometery/Geography—Visual Unity. Solstice: An Electronic Journal of Geography and Mathematics. Volume XVIII, No. 2. Ann Arbor: Institute of Mathematical Geography. http://www-personal.umich.edu/%7Ecopyrght/image/solstice/win07/hyperbolicgeometry.html*

for other centers of projection within the sphere and their counterparts outside the sphere. Thus, the entire unbounded class of perspective projections becomes bounded and subject to its own principle of duality. The infinite and unbounded set of centers of projection is now completely characterized as the bounded set within the sphere.

In the next section, you will have the opportunity to practice the concepts that have been discussed. Beginning with the geographic analysis of a world population data set, you will consider selected map projection issues from this chapter, but also keep in mind themes that have permeated this entire book, including data sources and scale.

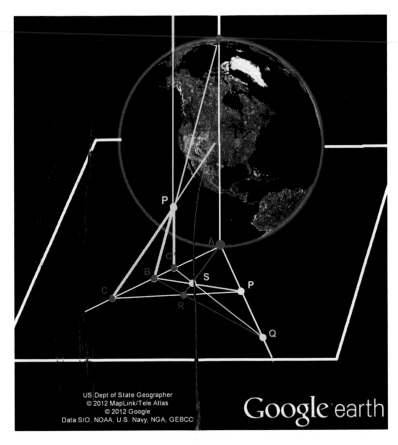

Figure 10.10 *Project P from the center of the sphere into the plane (gnomonic projection): P projects to C. Project P from infinity into the plane (orthographic projection): P projects to C'. Source: Arlinghaus, S. L. 2007. Geometery/Geography—Visual Unity. Solstice: An Electronic Journal of Geography and Mathematics. Volume XVIII, No. 2. Ann Arbor: Institute of Mathematical Geography. http://www-personal.umich. edu/~copyrght/image/solstice/win07/earth5.mov*

10.4 Practice using selected concepts from the chapter

10.4.1 Examining population change using the gridded population data set

Access a web browser and compare population change using the following resource from the Socioeconomic Data and Applications Center from Columbia University: http://sedac.ciesin.columbia.edu/gpw/. What patterns of density of persons per square kilometer do you notice around the planet? What are the reasons for the patterns? The density is far from uniform. East and South Asia stand out as the most densely populated areas, with other notably dense areas in west-central Africa, Central America, eastern North America, and Europe. Australia, the northern latitudes of Asia and North America, and north Africa stand out as the least densely populated. The reasons for the patterns are many, but they include historical settlement, climate, sustainable agriculture, ocean and land trading routes, and terrain.

Click on the 1990 population data layer and compare the 1990 data to the 2015 estimates, first by examining the region along the border between France and Germany from 1990 through 2015, and then northern Nigeria for the same time period. What demographic variables account for the larger increase in Nigeria versus that of France? The demography of Nigeria includes a much younger population and a correspondingly lower median age than that of France, resulting in a higher growth rate. Pan and zoom the map back to the France–Germany border. Do you think the population density really is that much higher in Germany versus that of France, or is this just an artifact due to the larger aggregations of the data collection units in Germany? As we have stated throughout this book, be critical of the data. Just because the data are shown on a single map does not mean that all the data came from the same source. Indeed, the Gridded Population Map of the World is highly dependent on the size of the administrative unit that was used as an input.

According to a variety of outside sources, Germany does have roughly twice the population density as does France (235 people per square kilometer for Germany versus 108 for France). But the map seems to indicate that broad areas of Germany are even higher in density, approaching that of the Netherlands. Is this accurate? Upon further investigation into the metadata, indeed, the resolution of the data set was to blame. CIESIN uses the highest-resolution data available for generating maps like this, and the data they were able to obtain for France had a much higher resolution. The resolution is calculated as the square root of the land area divided by the number of administrative units. For Germany, it was 28, and for France, it was four. The resulting population per administrative unit in 2000 was 184,000 for Germany versus only 2000 for France. Fortunately, CIESIN does an extremely good job documenting their sources and methods. The user still needs to make it a point to read that documentation. But what should we do when working with sites that do not document their data well? In today's world of myriad data, maps, and tools, it is more important than ever to have a good grounding in map interpretation and spatial analysis, but also to ask questions of the data you are using.

Under the map, use the population estimation service. Draw a polygon on the resulting map and obtain the total area and population for the polygon you drew. On the basis of the information in this chapter describing how the data were generated, how much confidence can you place in the population estimate? Do you think that the estimates vary in quality depending on where in the world you are drawing the box? Why or why not? Based on the above discussion of the France and Germany situation, the estimates do indeed vary in quality and input sources depending on where you are investigating.

Under the population map with the 1990 through 2015 data, access the "SEDAC Map Client" or go directly to the map on http://sedac.ciesin.columbia. edu/maps/client. Choose themes from the drop-down menu and compare variables of your own choosing to the population data. What three variables do you believe most directly influence population change around the world, and why? What three environmental and other variables do you believe are most directly impacted by rapid population growth and sustained high population density? Your answers may vary, but variables that contribute to rapid population growth include a high birth rate, a low death rate or a death rate that is lower than the birth rate, immigration that exceeds emigration, good health indicators, lack of deaths by natural hazards, major diseases, and political instability, and agriculture or imports that can support population growth.

10.4.2 Network analysis: Offline and online

10.4.2.1 Offline

Most of us were amazed when MapQuest gave us turn-by-turn directions back in the mid-1990s on the web. Nowadays, routing is not only ubiquitous in consumer mapping applications but is also embedded in thousands of organizations around the planet, from delivery services to tracking ships. It has become so familiar and expected that the complex computations requiring mathematics, geography, and GIS are usually overlooked. When we find a "route" we are looking for a connected path. That path might be analyzed in a number of ways. Is it unique? Is it the shortest path? Is there, in general, more than one shortest path (geodesic) in the given underlying geometry? Routing depends not only on the accuracy of the railroads, roads, shipping lanes, airline flight paths, and other mapped data, but also has come to depend on real-time information inputs, such as traffic, construction, and weather. Routing is also linked to and depends upon connectivity. Connectivity is a broad topic.

In a mapping context, consider a subway map in which the scale and track curviness are disregarded. The pattern of stops in correct relation to each other, along unrealistically straight track, suffices for most commuters. Curved lines are transformed into straight lines in this example often seen in large cities. What is dominant is connectivity. In that context, you will now have the opportunity to use various routing services in hands-on activities in the next section. These include determining the shortest route for emergency vehicles and the effect of traffic and barriers on travel time, distance, and route.

10.4.2.2 Online

Start the "real time closest facility" application on: http://nadev.arcgis.com/arcgis/samples/101/FlexProjects/CF-RT10/bin-release/Main.html. Once the application loads, click on a point anywhere on the map to start the response of fire and police vehicles. Note the routes the vehicles take to the incident, the relationship of distance and major roads to the incident, and the critical role that the ability to map and monitor real-time vehicle and incident locations makes in emergency response.

To further explore the effect that posted road speed and road width have on the travel time of specific vehicles, visit the Fast Service Area application on: http://nadev.arcgis.com/arcgis/samples/101/JS/FastServiceArea.html. This application shows the areas within a 5 minute drive time of the point your computer mouse lies on. Pan the mouse just north of Whittier in the mountains. Then pan the mouse to the northwest of this point, at the intersection of Highway 60 and I-605, noting the enlarged shape when the highways are included.

What influence does traffic have on accessibility and response? To find out, access the "routing with traffic" application on: http://nadev.arcgis.com/arcgis/samples/101/JS/Route101-traffic.html. Create a route on the map by clicking once for the starting point (move to a different location) and then by clicking another time for the ending point. Note how the time to travel that route changes based on traffic on US Highway 101. Increased traffic during peak commuting times definitely raises the time required to get from Point A to Point B.

At times, barriers from construction, special events such as parades, traffic, or emergencies can affect routing. To observe the effect of these barriers, visit the route task example for barriers, on: http://nadev.arcgis.com/arcgis/samples/RouteTaskEx/Barriers.html. Click on the map of London on Fitzrovia north of the River Thames to start the route and on Spring Gardens south of the river to end the route. How does the route take into account the natural barrier of the river? The route seeks obvious crossing points at bridges.

This map shows the fastest route between the two points. Why is the fastest route oftentimes not the shortest route? The fastest route could be because of lower traffic, one-way streets, fewer traffic lights, wider roads, or other factors. Now, set up some barriers: Draw a few point barriers, line barriers, and polygon barriers on the map and note how much time and distance they add to the route.

In the next section, you have the opportunity to construct your own routes, first for a tour bus company at a local scale in Manhattan, and then for a trucking company at a national scale for the USA.

10.4.3 Routing exercise: Determining best route for a tour bus in Manhattan

In this activity (**Figure 10.11**), you will use an ArcGIS Online routing service within ArcGIS for Desktop to design the best route for a tour bus through New York City, subject to certain conditions. As other GIS-based tasks migrate up to the cloud, so too are many tasks that have traditionally been in the desktop world of GIS. One example of this is that ArcGIS Online now includes a collection of geocoding tasks and a routing service that supports point-to-point and optimized routing for North America and Europe.

Suppose you are the new owner of a double-decker, open-top Manhattan tour bus. Your job is to route your bus from St John the Divine Church, to Radio City, the New York Public Library, the Empire State Building, the House of Oldies in Greenwich Village, the Woolworth Building, the American Geographical Society on Wall Street, and return to the starting point.

To solve this problem, start ArcGIS for Desktop and access the ArcMap application. Add from ArcGIS Online, the World Street Map. Zoom and pan the map so you are only looking at Manhattan Island in New York City. Under Extensions, make sure the Network Analyst extension is turned on. On the

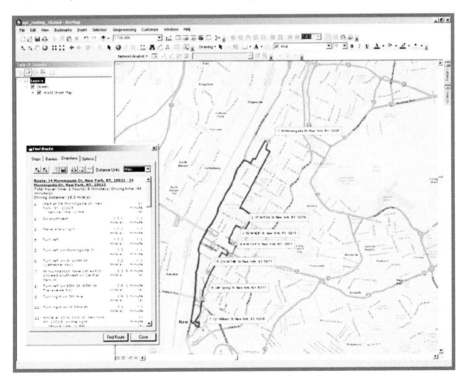

Figure 10.11 *New York City and Manhattan Tour Bus Routing. Source of base map: Esri software.*

Tools toolbar, select Find Route. On the Find Route tool, under Options, select the North American Routing Service (ArcGIS Online). On the same Options screen, check the box "Name locations added with tool using nearest address" and choose the following address locator: 10.0 North America Geocode Service (ArcGIS Online). On the same Options screen, mark the "Add route to map as" option as your choice of a graphic or a shapefile, with the symbology set to a red line.

Still on the Find Route tool, under the Stops tab, use the arrow with the dot to add your 7 stops as identified above. You might need to consult another data and map source to determine the locations of these seven landmarks in New York City. At the bottom of the window, select "Return to" (the first stop) so that you have a round trip. Then, select Find Route. Describe the results. Why do you suppose the return route followed the highway adjacent to the Hudson River instead of going back the same way? It did so because the highway is the fastest way through Manhattan, being adjacent to the river, the number of traffic lights is markedly fewer, there are a greater number of lanes, and the posted speed limit is higher. Next, experiment with the options. How does adding one stop, changing the order of stops, adding a barrier, adding a time window, or changing the impedance from "time" to "length" affect the final route? In **Figure 10.11**, it was estimated that 10 minutes would be required at each stop. How did this affect the overall time? No matter where you added the stops, each stop adds to the total time required. What kinds of mathematics are behind these accurate geocoding and routing services? The mathematics are complex, considering the attributes on each street segment, including average and posted speeds, one-way versus two-way streets, number of lanes, traffic, and time of day. Some of these attributes need to be included in the geodatabase before the analysis, and some can be added or adjusted by the data analyst as the analysis is being conducted. Either way, as we have seen throughout this book, data quality matters. Do you see why the "Manhattan" distance does not have unique geodesics? In a typical grid pattern there are generally two shortest routes to go from the southwest corner of a block to the northeast corner of the same block. The distances here are all based on large scale data, and map projection differences at large scales are minimal.

10.4.4 Routing exercise: Determining best route for trucking goods across the USA

Now that you have had some practice with a local routing activity, you next have the opportunity to perform some analysis at a national scale for the USA (**Figure 10.12**). Consider the following scenario: The Geo-Trucking Corporation, which delivers handheld mobile devices for field collection, needs to hire someone to drive a truck each week from Boston to Boise, and return, with intermediate stops in Birmingham, Alabama and Fort Smith, Arkansas. The Geo-Trucking Corporation has hired you to determine the most efficient route for this weekly trucking operation. To complete your task, you will use

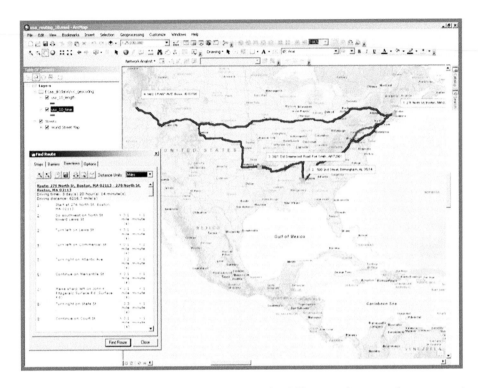

Figure 10.12 *Trucking activity map showing the differences between the route opti-mized by time versus the route optimized by length. Source of base map: Esri software.*

the same ArcGIS Online routing service that you used for the New York City tour bus activity to design the best route for this truck delivery service.

To solve this problem, start ArcGIS for Desktop and access the ArcMap application. Add from ArcGIS Online, the World Street Map. Zoom and pan the map so you are looking at the continental USA. Under Extensions, make sure the Network Analyst extension is turned on.

Since you are now working with national-scale data, you will need to make sure your routing service is optimized for the many more choices on routes that your trucking company can choose. To do so, you will make a server connection to the service that allows this optimization. Access the Windows drop-down menu and access "Catalog." In Catalog, expand "GIS Servers," Add ArcGIS Server, and then click Next. For your server URL, enter http://tasks. arcgisonline.com/arcgis/services, and then click Finish. On the Tools toolbar, select "Find Route." Click the Options tab, then click the Routing Service browse button. Click Server Data, then expand the tasks.arcgisonline.com server connection. Expand Network Analysis, and select ESRI_Route_NA, and click Open. Click the Routing Service dropdown, then select "Long_Route." On this same Options screen, check the box "Name locations added with tool using nearest address" and choose the following address locator: "10.0 US Streets Geocode Service (ArcGIS Online)." Change the impedance to "length." On the same Options screen, mark the "Add route to map as" option as your choice of a graphic or a shapefile, with the symbology set to a red line.

Still on the Find Route tool, under the Stops tab, this time, instead of using the arrow to add dots, use the "Add Stop" tool. Enter the following addresses, one at a time, and after adding each, right click and "add as stop in route": 276 North St., Boston, MA 02113; 1500 2nd Street, Birmingham, AL 35214; 3801 Old Greenwood Road, Fort Smith, AR 72903; and, 1402 Grant Ave, Boise, ID 83706. Once these points are loaded as stops, select Find Route. Describe the results. Remember that you chose, in this problem, to minimize the length. How did this choice affect the number of US and state highways that your truck will travel on versus interstate highways? Minimizing the length should allow for some US and state highways to be used, and not only interstate highways, since the goal here is to minimize the total distance the truck has to travel.

Next, experiment with the routing options. How does adding one stop, changing the order of stops, adding a barrier, or adding a time window affect the length and time of the resulting route? Any of these items should alter the total length and time. Repeat the process above, and this time, select the quickest route rather than the shortest route by changing the impedance from "length" to "time." Is there any difference? Why? There should be a difference because in analogy with the local example of New York City, the shortest route in distance is not necessarily the same as the fastest route in time. The map does show a difference, but it is only visible by zooming into larger scales. The route that is shortest in time occasionally sends your truck over different highways than it does for the route that is shortest in length. Around New York City, for example, your truck driver would actually save time to travel around New York City (the red road) instead of the traveling the shortest length route, which would go straight through New York City (the dark red road) (**Figure 10.13**).

To empirically discover the difference between shortest versus quickest route, save the routes as geodatabase feature classes or shapefiles. Then, edit the resulting tables in those geodatabase feature classes or shapefiles, populating the distance and time fields with the data that the routing tool combined into a description field. Summarizing these fields yielded 5999 miles and 98 hours for the shortest route, versus 6016 miles and 93.9 hours for the quickest route. Both routes could indeed be done in a week; the quickest route adds 17 miles but cuts over 4 hours of travel time. What factors might delay the total time required for your shipment? These factors could include rest stops for your driver, since you cannot expect any single driver to drive nearly 100 hours without sleep or breaks. Other factors causing delays could be traffic, stopping for toll booths, road construction, and so on. Through creating and comparing routes, activities such as this illustrate the spatial and mathematical thinking that is critical in the investigation of real-world problems.

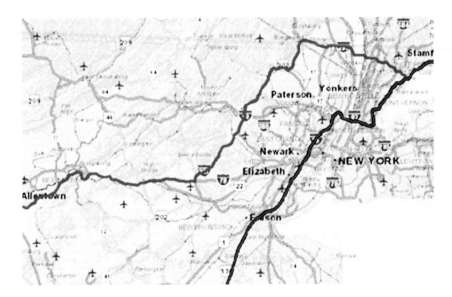

Figure 10.13 *Trucking activity, a closer look showing difference between the fastest route (in red) versus the shortest route (in dark red). Source of base map: Esri software.*

In the next activity, you will have the opportunity to conduct a deeper investigation, in particular, considering the type of traffic on streets, in your quest to locate the optimal site for a high speed Internet café in Denver, Colorado.

10.4.5 Find the busy streets—Denver

In **Chapter 3**, you created a map called "InternetCafeSites" as part of your investigation on siting the optimal location for your high-speed Internet café, using Census data. The activity below draws on that earlier map that you created. A link to the full study is provided below.

In **Chapter 3**, you found regions within 1 kilometer of a high school, university, or college—one component of solving the problem of where to locate the café in Denver. In **Chapter 7**, you found Denver neighborhoods in which the percentage of 18 to 21 year-olds is greater than 10% of the total population of those neighborhoods. Here we solve the remaining task of finding the busy streets in Denver.

Make a unique value map for roads, symbolizing the roads based on the attribute CFCC2 (Census Feature Class Code 2). Be sure to "add all values" to see all the possible values for CFCC2. Zoom in on the map when you are done, using the "I" identify button to answer the following questions: What is the CFCC2 code for the three most major road types in Denver City and County? In other words, these are the roads that carry the most traffic. What other information is in the roads table? The roads that are busy streets are those with a CFCC2 code of A1, A2, or A3.

10.4.6 Putting it all together: Practice—Denver Internet café activity

Now that you have all the pieces completed, you are close to determining the optimal site for your high speed Internet café in Denver. Putting those pieces together will draw on mathematics and geography. Use the Analysis Tools → Overlay → Intersect function to intersect areas within 1 kilometer of the schools, colleges, and universities together with the block groups containing at least 10% of their population between 18 to 21 years of age. Name your resulting intersection layer hs_univ_p1821 or something else that is suitable. How many polygons are under consideration? There should be 1797 polygons under consideration, but remember that these polygons are the result of a spatial intersection function. Therefore, the polygons do not represent block groups, but are small pieces from intersecting many disparate shapes, including your buffer zones and block groups. Near which two universities are these areas that are now under consideration for InstantWorld? The two universities under consideration are Regis University on the northwest side of Denver, and the University of Denver, in the south-central part of the city. The University of Colorado-Denver (Auraria) campus is the third major university in Denver. However, further investigation reveals that there are no neighborhoods nearby with a high percentage of 18 to 21 year-olds. Can you speculate why this might be the case? The reason is because this campus is a commuter-based school, and there are no dormitories in or near the campus that would largely house this age group.

It might be helpful for you to set a spatial bookmark at each of these locations under View → Bookmarks → Create. Now you have combined, and thought about, two of the three parts of the problem.

Recall that the final criterion is "on a busy street," with a CFCC2 code of A1, A2, or A3. Select these streets now. What expression did you use to select these streets? Based on the discussions in this book, you should know that you need to select these streets with an "OR" operation, as follows: CFCC2 = A1 OR CFCC2 = A2 OR CFCC2 = A3. How many street segments meet this "busy streets" criterion? The selection should result in 1134 out of 24,011 street segments selected.

The next selection that you need to perform is a spatial, not an attribute, selection. This will result in the selection of areas that are near schools and universities that also contain at least 10% of their population between 18 and 21 years old, that are also near busy streets. To do this, use the Select→ Select By Location function to intersect your hs_univ_p1821 layer (your "target" layer) with the busy streets (your "source" layer), as follows (**Figure 10.14**). Describe your result and indicate how many polygons are now selected in the hs_univ_p1821 layer. The result is a slightly smaller set of polygons in the

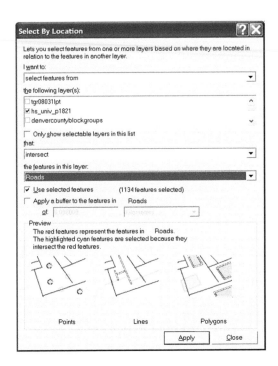

Figure 10.14 *Select By Location function. Source: Esri software.*

two neighborhoods under consideration; only those polygons near the busy streets, 1535 out of the original 1797 polygons are now selected. Export these selected polygons by right clicking on the layer, and using Data → Export into a data set named "internetcafesites" or another name that is suitable. Why could you not have used another Overlay→ Intersect operation to achieve the above results instead of the "select by location" function? This strategy would not have considered busy streets properly.

Now that you have narrowed your selection to just a few neighborhoods, you will note that there is one neighborhood where the median age (med_age) is the lowest (the "youngest" neighborhood) of the neighborhoods under consideration. You decide to make this neighborhood your #1 recommendation for the InstantWorld location. This selection is the area where the median age is 21.1, and it is located just south of Interstate Highway 25 and west of University Boulevard, on the north side of the University of Denver campus (**Figure 10.15**).

You used busy streets, median age, population 18 to 21 years old, and the location of colleges, universities, and high schools to determine the best location for your Internet café. What other criteria might you need to add to be able to make an even better decision to site this kind of business? How many of these criteria require knowledge about mathematics and geography? Other criteria that might be important include a map of zoning, vacant properties, properties for sale, location of competitors, price of land, traffic volume, location of

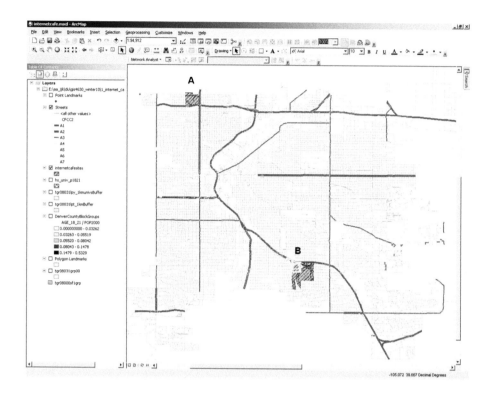

Figure 10.15 *Results of analysis for the optimal site for a high-speed Internet café in the City and County of Denver. The analysis resulted in two neighborhoods being under consideration, shown in red, near Regis University (A) and the University of Denver (B). The final site selected was the area with the youngest median age, near the University of Denver, shown in yellow. Source of base map: Esri software.*

bus routes, and so on. To take full advantage of considering all of these factors requires a keen knowledge of both mathematics and geography.

10.5 Graph theory and topology: Discrete and continuous spatial mathematics

The "practice" given above involves finding routes through geographic space and it draws on the intuition of the reader to understand what "routing" might mean, based on real-world experience. The science of "network analysis" and the mathematics of "graph theory" formalize and systematize ideas such as routing. They are broad and complex disciplines (Harary, 1969; Arlinghaus, Arlinghaus, and Harary, 2002).

A graph is a mathematical structure formed from a collection of nodes (vertices) and a collection of edges joining pairs of nodes. (Please do not confuse this sort of graph with graphs of functions.) The edges may show direction, as in the case of "going" or "coming back" on a trip; or they may not show direction. In the Manhattan tour bus example, the stops are nodes and the path of the bus indicates edges joining pairs of stops. The edges are directed, with direction on an edge corresponding to direction of the bus. Two consecutive stops are represented as nodes that are said to be adjacent. The edge linking these stops is said to be incident with each node. Graphs from graph theory are at the heart of "discrete mathematics" (as opposed to "continuous mathematics") and they are central in understanding many concepts in computer science as well as network analysis.

The graph-theoretic concept of connectivity was the basis for the routing activities. That concept is not only basic in graph theory but also in general topology. Because it is one central topic in both mathematical subfields, there are unfortunate instances in the published literature that make reference to "network topology" when what is meant in terms of mathematical terminology is "network connection pattern." Issues such as this one arise from difficulties traversing disciplinary boundaries.

As graph theory is a base for the study of discrete mathematics, general topology (Kelley, 1955; Bourbaki, 1966) is a base for the study of continuous mathematics. Topology studies properties preserved under continuous transformations (such as, for example, a stretching deformation). It is based on ideas that are fundamental to mapping as well as to geometry and set theory—ideas such as space, dimension, and transformation. Topology contains a number of subfields that one might study including point-set (general) topology, algebraic topology, and geometric topology. In application, one sees most often elements of topological spaces, from point-set topology, that involve connectedness, or compactness, or that involve a specialized topological space called a metric space. In this book, we have already drawn heavily on topological concepts beyond network connection pattern. The Jordan Curve Theorem is a topological theorem that has found application in geography in problems associated with street-addressing and geocoding. In our view of the Four-Color Theorem, another theorem from topology, we found application to map coloring and the associated communication of spatial content using color.

Indeed, the whole notion of transformation lies at the heart of both topology and spatial science. We considered stereographic projection from the North Pole of the sphere into a tangent plane at the South Pole of the sphere and saw that all points on the sphere, except the North Pole, map to that plane. We noted that the lack of mapping the projection pole into the plane was rooted in Euclidean geometry; in the non-Euclidean world, things can be different. In the inverse stereographic transformation, all points of the plane, map in a one-to-one fashion to the spherical surface, minus the North Pole. The two surfaces are topologically equivalent (homeomorphic). Add the missing point to the spherical surface (at the North Pole) and what had been a non-compact surface (a plane, of infinite extent) becomes compact. The topological concept of compactness underlies this transformation. This idea is the basis of the so-called "One-point Compactification Theorem" of topology (Alexandroff Extension, cited recently in 2011).

Do you see why this latter theorem explains that the same number of colors suffice for coloring maps on these two different surfaces? Consider any map on the surface of a sphere. Poke a hole in the interior of any region on the map on the spherical surface. Use that hole as the "North projection Pole" from which to project stereographically the map on the spherical surface into a plane tangent at the antipodal point to the North projection Pole. The map in the plane can be colored using x colors. Use inverse stereographic projection to pull the map in the plane back to the spherical surface. The entire map on the spherical surface, except the North projection Pole, is now colored. Color the missing point on the spherical surface with the same color as the region that surrounds it (hence the need for selecting an interior point as the North projection Pole). Thus, however many colors were needed in the plane is the same number as would be needed on the surface of a sphere. Does this result surprise you? Do you see that the surface of a sphere is, as is the plane, a two-dimensional surface? This result was known well in advance of proving the actual number needed; yet another instance of precision before accuracy!

10.6 Putting it all together: Theory

The issues that we face as a society in the twenty-first century, such as population change, sustainable agriculture, natural hazards, energy, water quality and availability, climate change, crime, and more, vary widely in scale and discipline. Yet they share several key characteristics: They are all complex issues that are of global importance and yet increasingly affect our everyday lives. They are all inherently geographic issues—they all have to do with "where," and more important, the "whys of where." They all depend heavily on mathematics—to represent, process, and manipulate data. They all rely on the kinds of critical thinking, inquiry, and problem-solving skills that we have emphasized in this book. As such, gaining skills in spatial mathematics will enable anyone to be a more effective decision maker. Given the issues outlined above, the integration of mathematics and geography will become even more important in the future.

We hope the reader has enjoyed seeing a small sample of what "spatial mathematics" has to offer. This book has presented classical materials, such as the material associated with Eratosthenes, it has presented contemporary materials such as the population potential maps of Grayson, and it suggests direction into the future in the use of non-Euclidean geometry in perspective map projection. The reader interested in learning about graph theory, network analysis, topology, and more, is referred to a number of references or may await future works in this series that make even more direct linkages between spatial mathematics and these fields.

By studying disparate spatial examples within a broad mathematical context, we enlarge our own capabilities to think logically and clearly. Our goal is that others feel that their intellectual horizons have been broadened, as well!

10.7 Related theory and practice: Access through QR codes

Theory

Persistent archive:

University of Michigan Library Deep Blue: http://deepblue.lib.umich.edu/handle/2027.42/58219

From Institute of Mathematical Geography site: http://www.imagenet.org/

Arlinghaus, S. L. 2010. Fractals Take A Non-Euclidean Central Place. *Solstice* Volume XXI, No. 1. http://www.mylovedone. com/image/solstice/sum10/HyperbolicCentralPlaceFractals.html

Arlinghaus, S. L. 2009. Essays on Mathematical Geography: Contemporary Visualizations. *Solstice* Volume XX, No. 2. http:// www.mylovedone.com/image/solstice/win09/Arlinghaus.html

Arlinghaus, S. L. 2007. The Animated Pascal. *Solstice* Volume XVIII, No. 2. http://www-personal.umich.edu/~copyrght/ image/solstice/win07/MacLaurin.html

Arlinghaus, S. L. 2007. Desargues's Two-Triangle Theorem. *Solstice* Volume XVIII, No. 2. http://www-personal.umich. edu/~copyrght/image/solstice/win07/Desargues.html

Arlinghaus, S., M. Batty, and J. Nystuen. 2003. Animated Time Lines: Coordination of Spatial and Temporal Information. *Solstice* Volume XIV, No. 1. http://www-personal.umich.edu/~copyrght/image/solstice/sum03/batty.html

Arlinghaus, S. L., W. C. Arlinghaus, and F. Harary. 2002. *Graph Theory and Geography: An Interactive View eBook.* John Wiley and Sons, New York. Archived in Deep Blue. http://deepblue.lib.umich.edu/handle/2027.42/58623

Arlinghaus, S. 1986. The Well-tempered Map Projection. *Essays on Mathematical Geography.* Monograph #3. Ann Arbor: Institute of Mathematical Geography, pp. 1-2. "Books" http://www.imagenet.org/

Practice

From Esri site: http://edcommunity.esri.com/arclessons/arclessons.cfm

Kerski, J. Siting an Internet Café in Denver Using GIS. Complexity level: 3. http://www.mylovedone.com/Kerski/Denver.pdf

Kerski, J. Routing Using ArcGIS Online. Complexity level: 4. http://edcommunity.esri.com/arclessons/lesson.cfm?id=441

Kerski, J. Siting a Ski Area: Traffic Volume Complexity level: 3, 4, 5. http://edcommunity.esri.com/arclessons/lesson. cfm?id=627

Glossary

Imagination is more important than knowledge.
For knowledge is limited, whereas imagination embraces the entire world,
stimulating progress, giving birth to evolution.
It is, strictly speaking, a real factor in scientific research.

Albert Einstein

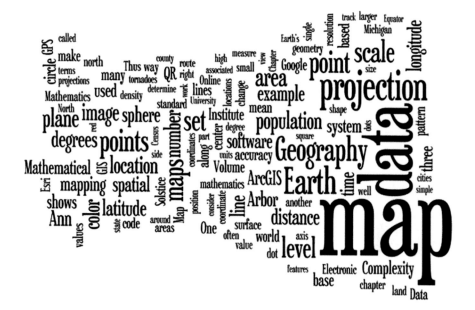

Word Frequency counts from Word Cloud software, Wordle.net: Map 470, Data 315, Projection 168, Earth 148, Geography 133, Point 123, Scale 121, Set 119, Level 113, Area 106, Plane 101, Population 97, Location 94, Image 94, Color 92, Degrees 92, Number 91, Latitude 89, Distance 89, Ann 87, Line 86, Sphere 85, Longitude 84, Spatial 84, Arbor 83, Software 82, Time 80, ArcGIS 80, Mapping 80, System 76, Mathematical 75, Complexity 72, Circle 70, Base 70, QR 69, Accuracy 65, World 65, GPS 64, Volume 64, Mean 63, Center 62, Pattern 60, GIS 59, Code 59, Values 57, Dot 56, Surface 56, Mathematics 55, Standard 51, Land 51, Online 50, Geometry 50, Resolution 49, Coordinate 48, Route 47, Density 47, Michigan 47, Axis 46, Position 46, Tornadoes 44, Shape 44, State 43, Cities 43, Measure 43, Census 41, Units 41, County 41, Track 40, View 40, Equator 40, Features 39, Hexagons 37, Block 37, Method 37, Distortion 37, Denver 37, Parallel 37, Globe 37, Circumference 36, Theory 36,

Buffers 35, Analysis 35, Local 35, Direction 35, Length 34, Place 34, Polygons 34, Function 34.

Knowledge, however, may (in part) feed the imagination. To that end, we offer a set of terms and definitions or comments about them. Some are technical, others are common words used with perhaps uncommon meaning within this context. The word cloud for the entire book selects words based only on frequency. In contrast, in this glossary the authors select approximately ten words or short phrases from each of the 10 chapters and offer brief explanatory material about each of them.

Absolute location: Exact location designated using a recognized coordinate system.

Accuracy: A measure of whether points taken in a system are close to an accepted value.

Antipodal points: Points at opposite ends of a diameter of a sphere.

Aspect ratio: Ratio of horizontal dimension to vertical dimension (or vice-versa).

Attribute table: In a GIS, a table of data linked to a map containing characteristics of the data being mapped.

Azimuthal: In mapping, the projection of the surface of the globe to a plane.

Bar scale: A graphical way of expressing map scale that changes with enlargement and reduction of the map.

Base map: The first layer of a map; a map to build from. Often a simple outline map. A GIS, often refers to the raster layer such as imagery or topography that is typically streamed online.

Blocks: A statistical area for Census purposes typically bounded by city streets or some other physical feature. Blocks are aggregated into Block groups; blocks are usually synonymous in urban areas to city blocks.

Block groups: A cluster of blocks that serves as a statistical collection unit for US Census data.

Buffer: In a GIS, the zone around a map feature measured in units of time or distance; used in proximity analysis.

Cartesian coordinate system: Specifies any given point in the plane uniquely by a pair of numerical coordinates which are signed distances from the point to two fixed perpendicular lines that intersect at the origin; named for René Descartes.

Central tendency: A single summary measure for one variable—an average.

Choropleth map: A thematic map with areas shaded in proportion to the measurement of the variable being mapped.

Classification: In mapping, the partitioning of data according to various systematic methods. Quantile, natural breaks, geometrical interval, equal interval, standard deviation are common methods.

Close-packing: Packed tightly together; compactly arranged.

Cluster: Grouping of data.

Commensurable: Measureable by a common standard.

Conformal: In mapping, where the shape of a small area of the surface mapped is preserved; preserving local angles.

Conic: In mapping, where the surface of the globe is projected onto the surface of a cone.

Connectivity: A pattern of connection. May have specific meanings in various contexts: in graph theory, in topology, in communications, in brain mapping and other medical contexts, and more.

Contour: An isoline based on elevation data, or height above a fixed reference point.

Cylindrical: In mapping, where the surface of the globe is projected onto the surface of a cylinder.

Datum (pl.: datums): Reference from which measurements are made. In mapping, the datum reference points are based on a mathematically calculated shape of the Earth.

Decimal degrees: A format of absolute location where the latitude and longitude values are expressed as decimal fractions.

Degree Confluence Project: An effort to document absolute locations on the Earth through narratives and photographs using whole-degree intersections of latitude and longitude.

Demographic analysis: Measurement of the dimensions and dynamics of a population.

Digitizing: In GIS, the process of converting paper map features into a digital format.

Directional trend: Measures directional influences on data sets.

Dispersion: The spread of a scatter of dots—contrast with clustering.

Dot density map: A thematic map that displays information by showing numerous dots, each of which represents some specific quantity.

Elevation: The height of a geographic location above a fixed reference point.

Equal area projection: A map of constant area scale. Visual comparisons of one area to another are true. On an equal area projection Greenland does not appear larger than Brazil.

Equidistant: A map projection that preserves distance from a standard point or line.

Four Color Theorem: Four colors suffice to color a map in the plane (with adjacency defined across edges, not corner points, of polygons).

Fujita scale: A scale for rating the intensity of tornadoes.

Function: A mathematical mapping such that each element of a given set (domain) is associated with an element of another set (range).

Geocoding: The process of assigning spatial coordinates to geographic features in a GIS. Generally, it involves interpolating spatial locations from street addresses or other spatially referenced data sets.

Geodesic: Loosely, the shortest path on a surface.

Geographic Center: The point on a map that represents the centroid of the geographical shape defining it. Specifically, it is the point in an area at the intersection of all straight lines that divide that area into two parts of equal moment about the line, or the arithmetic mean of all points of the shape.

Geoid: A mathematically-computed surface of the Earth based on constant gravity, used in computing absolute location on the Earth.

Geotagging: The process of adding spatial data (often coordinates) to a photograph or other media.

GIS: Geographic Information System—Software combining cartography, statistical analysis, and database technology to map data to enable users to visualize spatial patterns. More broadly, GIScience, or Geographic Information Science.

GPS: Global Positioning System—An accurate global navigational and surveying system based on an array of satellites and the signals from those satellites as received by on-the-ground devices.

Graph theory: Mathematics behind much network analysis; used to model pairwise relations between elements of a given set. It is a foundational branch of discrete mathematics.

Graticule: In mapping, a grid system, such as latitude–longitude, displayed on a sphere.

Great circle: A circle formed by the intersection of the surface of a sphere with a plane passing through the center of that sphere.

Hexagon: A six-sided polygon.

Hierarchy: In mathematics, an ordered set.

Hue: A color's direction from white in a color wheel.

Isoline: A line on a map joining points of equal value.

Join: In a GIS, to establish a relationship between a non-mapped table and a map layer's attribute table. The relationship may be one-to-one or many-to-one.

Jordan Curve Theorem: A topological theorem to systematically identify inside and outside of simple closed curves. In mapping it has implications for digitizing and geocoding.

Latitude: An angular measure on the globe indicating distance North or South of the Equator, often expressed in degrees and minutes.

Lattice: In mathematics, a partially-ordered set.

Longitude: An angular measure on the globe indicating distance East or West of the Prime Meridian, often expressed in degrees and minutes.

Luminosity: The brightness of a color; how light or dark it is.

Manhattan space: Geometry that roughly parallels the pattern of the street grid of Manhattan; geodesics are not unique.

Map: A representation of data in a spatial context.

Mean center: In geography, the point determined from the average x and y values for a scatter of input points.

Meridian: A half of a great circle on the globe joining the North and South poles.

Military Grid Reference System: A system to identify locations on the Earth formed from two sets of parallel lines intersecting at right angles and forming a grid of squares.

Nested hierarchy: A set of layers of polygons which fit neatly in association with each other as one moves between layers.

Nested polygons: A system wherein smaller polygons fit together within larger ones with no gaps and with no area spilling outside the larger ones.

Network analysis: The systematic study of geographical (or other) networks, often using mathematics designed for the purpose.

Non-Euclidean geometry: Geometries that do not obey Euclid's Parallel Postulate.

Normalizing data: In mapping, dividing one variable by another to create ratio maps; for example, dividing population by state by total national population in order to look at the ratio measure of population density.

Obelisk: A stone pillar, often a landmark, with a rectangular cross-section and a pyramidal top.

Oblate spheroid: An ellipsoid generated by rotating an ellipse about one of its axes such that the diameter of the equatorial circle is greater than the length of the axis of revolution. The Earth is an oblate spheroid with a polar axis shorter than its equatorial diameter.

Parallel: A small circle on the globe formed by a plane parallel to the equatorial plane.

Parallel Postulate (Euclid): In two-dimensional geometry, given a line m and a point P not on m. There is exactly one line passing through P that does not intersect m (is parallel to m).

Partitioning of data: The separation of data into mutually exclusive and exhaustive classes.

Perspective projection: In mapping, the projection of the globe to a surface as seen through a single point.

Pixel/voxel: A pixel is a picture element—a basic unit on a computer screen. A voxel is a 3D (volume) pixel.

Population potential: The accessibility of people from a given point—how close people are to a point.

Precision: A measure of whether points taken within a system are close to each other.

Prime number: A natural number that has exactly two natural number divisors, itself and 1.

Projection: In mapping, the process of sending points on the globe to points in the plane, thereby flattening the Earth's surface and allowing data to be displayed in two dimensions on a paper or digital map.

Proximity: Nearness measured in terms of place, time, order, or other concepts.

Public Land Survey System: Mathematically designed system as a nationally conducted cadastral survey.

QR code: Quick Response code. A three-dimensional bar code.

Quantile: A method of data classification dividing the classes such that each class contains the same number of observations.

Randomizing layer: In a dot density map, the layer in which the scatter of dots is initially set forth.

Raster format: A data structure composed of a rectangular grid of pixels.

Reference ellipsoid: A mathematically-defined shape that is a "best-fit" to the geoid, most often for a continent, but potentially for the entire globe. Smoother than the geoid, it is the surface upon which coordinate systems are defined.

Regular/irregular data pattern: The evaluation of pattern that invokes concepts such as symmetry or its lack thereof.

Relative location: Referencing positions on the Earth by indicating displacement from another location or site, which may include terms such as near, adjacent, left, or ahead.

Remote Sensing: The science of acquiring and interpreting imagery collected from indirect recording of a specific band or bands of energy reflected from the Earth, such as satellite imagery.

Representative fraction: A fraction expressing map scale that is independent of units.

Resolution: Sharpness or level of detail in an image. It is directly related to the number and density of pixels.

RGB/CMYK: Color models by which all colors are described as a mixture of base colors. RGB—Red, Green, Blue; CMYK—Cyan, Magenta, Yellow, and Black.

Routing: The planning of a line through a set of points (route) through real or imaginary locations.

Saturation: The degree to which a color is intense or concentrated.

Scale: A numerical relationship, expressed as a ratio, between the actual size of an object and the size of an image that represents it on a map, plan, or diagram. A map scale is an expression of how many units on the map represent a corresponding number of units on the Earth's surface. The manner of expression may take various forms.

Set intersection/union: The intersection of sets A and B is composed of all elements simultaneously in both sets and no other elements. Union of sets A and B is all elements in A together with all elements of B—in A or B or both.

Shaded relief: Hill shading or coloring to simulate the shadow on a raised relief map.

Small circle: A circle formed by the intersection of the surface of a sphere with a plane intersecting the sphere at more than a single point but not passing through the center of the sphere.

Spatial Analysis: With GIS, analyzing geographic phenomena in terms of their distribution on a map. It might involve mapping elements of slope and aspect, attributes, connection and network patterns, hydrological models, overlays, geocoding, and others.

Spider diagrams: Conceptual visualizations of regions and connections between regions.

Standard deviational ellipse: An area representing one standard deviation from the mean center of all the data in a data set. It measures whether mapped features exhibit a directional trend and the amount of scatter of those features.

State Plane Coordinate System: A set of 124 geographic zones or coordinate systems created to refer to ground positions in the United States of America.

Stereographic projection: A perspective projection of the globe from a point on the surface to a plane tangent at the point antipodal to the selected surface point of projection.

Street addressing: A method of assigning a numerical designation to properties with regards to their position in the transportation network; may vary regionally and from country to country.

Symbology: A system of (map) symbols, including size, color, shape, and other attributes.

Tessellation: A pattern of shapes that fit together with no gaps; a tiled floor is a physical model.

Thiessen polygon: Individual area of influence around each point in a distribution.

TIGERweb: Interactive online map of US Census data.

Tissot Indicatrix: In mapping, the use of small ellipses to measure projection distortion.

Topographic map: A map that indicates, through a series of contour lines or hill shadings, the relief or physical characteristics of a landscape. A topographic map also typically shows other physical features such as rivers and human-constructed objects such as buildings and roadways.

Topology: A branch of mathematics concerned with properties that are preserved under continuous transformations.

Tracts or Census tracts: A geographic region defined for the purpose of the taking of a census. It is composed of a group of block groups, typically containing a few thousand residents. Census tracts nest within county equivalents.

Transformation: A mathematical mapping from one set to another (or itself).

UTM: Universal Transverse Mercator—A grid-based method of designating absolute locations on the surface of the Earth.

Vector format: A format for storing spatial data as points, lines, or areas (polygons).

Voxel: A voxel is a 3D (volume) pixel.

Web GIS: Geographic Information Systems tools, maps, and data served on the Internet; also sometimes referred to as web mapping.

Web Mercator Auxiliary Sphere: A tiling scheme to produce sphere-based geographic coordinate system.

Weighted mean center: In a GIS, the point considered to be the center of a distribution of points with weight added in relation to some attribute, such as population, pH, or noise level.

References, Further Reading, and Related Materials

These materials are in addition to materials at the end of each chapter; those focus on documents created by the authors of this work.

Abbot, E. A. 1884. *Flatland: A Romance of Many Dimensions*. United Kingdom: Seely & Co.

Adam, J. A. 2012. *X and the City: Modeling Aspects of Urban Life*. Princeton: Princeton University Press.

Aitchison, A. 2011. The Google Maps / Bing Maps Spherical Mercator Projection http://alastaira. wordpress.com/2011/01/23/the-google-maps-bing-maps-spherical-mercator-projection/

Albrecht, J. 2005. Maps projections. Hunter College: Introduction to Mapping Sciences. http://www.geo.hunter.cuny.edu/~jochen/gtech201/lectures/lec6concepts/map%20 coordinate%20systems/how%20to%20choose%20a%20projection.htm

Alexandroff Extension. 2011. http://en.wikipedia.org/wiki/Alexandroff_extension

Anderson, J. R., E. E. Hardy, J. T. Roach, and R. E. Witmer. 1976. A Land Use and Land Cover Classification System for Use with Remote Sensor Data. *US Geological Survey Professional Paper* 964, 41 pages.

Appel, K. and W. Haken. 1976. A proof of the 4-color theorem. *Discrete Mathematics*, 16, no. 2, 179–180.

Arlinghaus, S. L. 2007. Geometry/Geography—Visual Unity. *Solstice: An Electronic Journal of Geography and Mathematics*. Volume XVIII, No. 2. Ann Arbor: Institute of Mathematical Geography. http://www-personal.umich.edu/~copyrght/image/solstice/win07/hyper bolicgeometry.html

Arlinghaus, S. L. et al. 1994. *Practical Handbook of Curve Fitting*. Boca Raton: CRC Press.

Arlinghaus, S. L. 1993. Central Place Fractals. Chapter 10 in *Fractals in Geography*, edited by N. Lam and L. DeCola. New Jersey: Prentice-Hall.

Arlinghaus, S. L. 1993. Electronic Geometry. *The Geographical Review* April, Vol. 83, No. 2, 160–169.

Arlinghaus, S. L. 1985. Fractals take a central place. *Geografiska Annaler*, Journal of the Stockholm School of Economics, 67B, 83–88.

Arlinghaus, S. L. and W. C. Arlinghaus. 1989. The fractal theory of central place hierarchies: A Diophantine analysis of fractal generators for arbitrary Löschian numbers. *Geographical Analysis: An International Journal of Theoretical Geography*. Ohio State University Press. Vol. 21, No. 2, April, pp. 103–121.

Arlinghaus, S. L., W. C. Arlinghaus, and J. D. Nystuen. 1990. The Hedetniemi Matrix Sum: An Algorithm for Shortest Path and Shortest Distance, *Geographical Analysis*, Vol. 22, No. 4, 351–360.

Arlinghaus, W. E. 2011. Personal communication.

Arnold, D. N. and J. Rogers. 2007. Möbius Transformations Revealed. http://www.youtube. com/watch?v=JX3VmDgiFnY

Barbaree, D. Watsons Go To Birmingham with AGXO. Complexity level: 2, 3. http:// edcommunity.esri.com/arclessons/lesson.cfm?id=643

Bender, B. 1999. Subverting the Western gaze: Mapping alternative worlds. In P.J. Ucko and R. Layton. *The Archaeology and Anthropology of Landscape: Shaping your landscape*. One World Archaeology. **30**. London: Routledge.

Berry, J. 2002. Maps Are Numbers First, Pictures Later. *GeoWorld*, August 2002, pp. 20–21. http://www.innovativegis.com/basis/mapanalysis/Topic18/Topic18.htm#Numbers_first_pictures later

Birkhoff, G. and Mac Lane, S. 1953. *A Survey of Modern Algebra*. Revised Edition. New York: Macmillan.

Bogue, Donald J. 1950. The structure of the metropolitan community: A study of dominance and subdominance. Ann Arbor: Horace H. Rackham School of Graduate Studies, The University of Michigan.

Bourbaki, N. 1966 *Topologie Générale (General Topology)*. Boston: Addison–Wesley.

Boys, C. V. 1902. *Soap Bubbles and the Forces which Mould Them*. London: Society for Promoting Christian Knowledge.

Branting, S. GPS and Heron's Algorithm. Complexity level: 3. http://edcommunity.esri.com/arclessons/lesson.cfm?id=541

Branting, S. The Wisdom in GPS Crowds. Complexity level: 3. http://edcommunity.esri.com/arclessons/lesson.cfm?id=456

Bret V. 2011. *Kill Math*. http://flowingdata.com/2011/10/05/ kill-math-makes-math-more-meaningful/

Campbell, J. F. and M. E. O'Kelly. 2012. Twenty-Five Years of Hub Location Research. *Transportation Science,* 46(2), 153–169. http://www.geography.osu.edu/faculty/okelly/journals.php?filter=H

Catawba County Building Services. 2012. North Carolina. QR code application. http://www.catawbacountync.gov/depts/u%26e/pdfs/hardcard201101.pdf

Christaller, W. 1933. *Die zentralen Orte in Süddeutschland.* Jena: Gustav Fischer. (Translated (in part), by Carlisle W. Baskin, as *Central Places in Southern Germany*. Prentice Hall 1966.)

Christaller, W. 1941. Struktur und Gestaltung der Zentralen Orte des Deutschen Ostens, Gemeinschaftswerk im Auftrage der Reichsarbeitsgemeinschaft für Raumforschung, Teil 1, Dr. Walter Christaller, *Die Zentralen Orte in den Ostgebieten und ihre Kultur-und Marktbereiche*, Leipzig: K. F. Koehler Verlag.

Clarke, K. C. and P. D. Teague, 1998. Cartographic symbolization of uncertainty. *Proceedings*, American Congress on Surveying and Mapping conference, http://www.geog.ucsb.edu/~kclarke/Papers/ACSM98.pdf.

Copernicus, N. 1543. *De revolutionibus orbium coelestium*.

Coulter, B. and Kerski, J. 2005. Using GIS to transform the mathematical landscape. *Technology-Supported Mathematical Learning Environments*: 67, Chapter 22. The Yearbook of the National Council of Teachers of Mathematics.

Coxeter, H. S. M. 1961. *Introduction to Geometry*. New York: John Wiley & Sons. 53–54.

Coxeter, H. S. M. *Non-Euclidean Geometry, 6th ed*. Washington, DC: Math. Assoc. Amer., 1998; earlier, 1965.

Coxeter, H. S. M. 1961. *The Real Projective Plane*. Cambridge: Cambridge University Press.

Coxeter, H. S. M. and S. L. Greitzer. 1967. *Geometry Revisited*. Washington, DC: Math. Assoc. Amer.

Dacey, M. F. 1965. The geometry of central place theory. *Geografiska Annaler*, B, 47, 111–124.

Dana, P. H. Map Projections. http://www.colorado.edu/geography/gcraft/notes/mapproj/mapproj_f.html.

DeBlij, H. 2008. *The Power of Place: Geography, Destiny, and Globalization's Rough Landscape*. Oxford: Oxford University Press.

Deetz, C. H. and O. S. Adams. 1934. *Elements of Map Projection with Applications to Map and Chart Construction, 4th ed*. Washington, DC: US Coast and Geodetic Survey Special Pub. 68.

Degree Confluence Project, accessed July 18, 2012. http://confluence.org/

Delport, Jason. Create QR code, Google Chart API. http://createqrcode.appspot.com/

de Smith, M., P. Longley, and M. Goodchild. 2006. *Geospatial Analysis—A comprehensive guide*. http://www.spatialanalysisonline.com/

Duke, B. X and Y Coordinates Rule the Map. Complexity level: 5. http://edcommunity.esri.com/arclessons/lesson.cfm?id=252

Einstein, A. 1919. Quotation found online in article in *Wikiquote*. http://en.wikiquote.org/wiki/Albert_Einstein

Eppstein, D. 2012. Geometry in Action. http://www.ics.uci.edu/~eppstein/gina/voronoi.html

ESRI. 2004. ESRI Cartography: *Capabilities and Trends*. Redlands, CA. White Paper.

ESRI references on Web Mercator Auxiliary Sphere. http://forums.arcgis.com/threads/8762-WGS-1984-Web-Mercator-(Auxiliary-Sphere)-WKID-102100

Faulkner, W. 1950. Nobel Prize Acceptance Speech. http://www.mcsr.olemiss.edu/~egjbp/faulkner/lib_nobel.html

Feinberg, J. 2011. *Wordle*. Website: http://www.wordle.net/

Fejes-Toth, L. 1968. Solid circle-packings and circle-coverings, *Studia Sci. Math. Hungar.* 3, 401–409.

Ferwerda, C. Issues of Scale: Environmental Justice and the MAUP. Complexity level: 4. http://edcommunity.esri.com/arclessons/lesson.cfm?id=716

Friedemann, S. GeoSetter. http://www.geosetter.de/en/

Gauss, C. F. 1801. Disquisitiones Arithmeticae.

Gauss, C. F. 1840. Untersuchungen über die Eigenschaften der positiven ternären quadratischen Formen. *Journal für die Reine und Angewandte Mathematik.* 20, 312–320.

Gershenson, Daniel E. 1964. *Anaxagoras and the Birth of Scientific Method*. New York: Blaisdell Publishing Company.

Google Earth http://www.earth.google.com/

Grayson, J. 2012. Population Potential map. http://maps.esri.com/AGSJS_Demos/PopulationPotential/

Gridded Population of the World (GPW) (http://sedac.ciesin.columbia.edu/gpw/). Columbia University's Center for International Earth Science Information Network (CIESIN). Originally produced by Waldo Tobler, Uwe Deichmann, Jan Gottsegen, and Kelley Maloy from the University of California Santa Barbara, with partial support from CIESIN under a US NASA grant.

Guenette, M. 2009. Triangulation of a Hierarchical Hexagon Mesh. M.S. Thesis, School of Computing, Queen's University, Kingston, Ontario, Canada. http://hdl.handle.net/1974/1665

Hage, P. and F. Harary. 1996. *Island Networks: Communication, Kinship, and Classification Structures in Oceania*. Cambridge: Cambridge University Press.

Haggett, P., A. D. Cliff, and A. Frey. 1977. *Locational Analysis in Human Geography 2: Locational Methods*. Second Edition. New York: John Wiley & Sons.

Hanebuth, E. SkillsUSA 2007 Geospatial Competition Exam. Complexity level: 5. http://edcommunity.esri.com/arclessons/lesson.cfm?id=327

Harary, F. 1969. Graph Theory. Boston: Addison-Wesley.

Hausdorff, F. 1914. *Grundzüge der Mengenlehre*. Leipzig, Germany: von Veit.

Herstein, I. N. 1964. *Topics in Algebra*. New York: Blaisdell.

Hilbert, D. and S. Cohn-Vossen. 1952. *Geometry and the Imagination*. Second edition. New York: Chelsea.

Imus, D. and P. Dunlavey. 2002. *Back to the Drawing Board: Cartography vs. the Digital Workflow*. MT. Hood, Oregon.

Internet Archive. 2005. *WayBackMachine: Map Projection Pages*. http://web.archive.org/web/20070104121606/http://www.3dsoftware.com/Cartography/USGS/MapProjections/

Jeans, J. H. 1929. *Eos, or, The Wider Aspects of Cosmogony*. New York: E. P. Dutton and Company.

Jeer, S. with B. Bain. 1997. Traditional color coding for land uses. *American Planning Association*. http://www.gsd.harvard.edu/gis/manual/style/ColorConventions.pdf

Jefferson, M. 1928. The Civilizing Rails. *Economic Geography*, 4, 217–231.

Johnson, W. E. 1907. *Mathematical Geography*. New York: American Book Company.

Jordan, C. 1909. *Cours d'Analyse*. Paris: Gauthier-Villars.

Juster, N. 1963. *The Dot and the Line*. New York: Random House.

Kelley, J. L. 1955. *General Topology*. New York: van Nostrand.

Kent, A. J. 2005. Aesthetics: A lost cause in cartographic theory? *The Cartographic Journal* 42 (2): 182–188.

Kepler, J. 1609. *Astronomia nova*.

Kerski, J. J. 2010. Pointed Journeys, *The American Surveyor*, Volume 7, Issue 2. http://www.amerisurv.com/PDF/TheAmericanSurveyor_Kerski-PointedJourneys_Vol7No2.pdf.

Kolars, J. F. and J. D. Nystuen. 1974. *Human Geography: Spatial Design in World Society*, Englewood Cliffs: McGraw-Hill. Original drawings by Derwin Bell.

Koldony, L. 2011. New York City to Put QR Codes On All Building Permits by 2013. http://techcrunch.com/2011/02/22/nyc-qr-codes-on-buildings/

Kopec, R. J. 1963. An alternative method for the construction of Thiessen polygons. *Professional Geographer*, 15 (5): 24–26.

Kraak, M.-J. and F. Ormeling. 2002. *Cartography: Visualization of Spatial Data*. Englewood Cliffs: Prentice Hall.

Krause, E. 1987. *Taxicab Geometry: An Adventure in Non-Euclidean Geometry*. New York: Dover.

Kuhn, T. S. 1957. *The Copernican Revolution: Planetary Astronomy in the Development of Western Thought*. Cambridge, MA: Harvard University Press.

Kuhn, T. S. 1962. *The Structure of Scientific Revolutions*. Chicago: University of Chicago Press.

Ladak, Alnoor and Martinez, Roberto B. Automated Derivation of High Accuracy Road Centrelines Thiessen Polygons Technique. http://proceedings.esri.com/library/userconf/proc96/to400/pap370/p370.htm

Lease, C. A. 1979. Practical Approach Counts In Math Class. Sunday, January 21, 1979. *Columbus Dispatch*. A-7.

Leibniz, G. W. 1686. De Geometria, *Acta Eruditorum*.

Lévi-Strauss, C. 1949. *The Elementary Structures of Kinship*.

Lévi-Strauss, C. 1969. *The Elementary Structures of Kinship*. Rev. ed. Trans. James Harle Bell, John Richard Stermer, and Rodney Needham. London: Eyere & Spottiswoode.

Lewin, K. 1936. *Principles of Topological Psychology*. Translated by F. Heider and G. M. Heider. New York: McGraw-Hill Book Company.

Loeb, Arthur L. 1976. *Space Structures: Their Harmony and Counterpoint*. Reading, MA: Addison-Wesley.

London School of Economics (LSE) Library. 2012. Swedish government information: Statistical data. Accessed July 18, 2012. http://www2.lse.ac.uk/library/collections/govtpub/sweden/sweden_government_statistics.aspx

Lösch, A. 1954. *The Economics of Location*, translated by William H. Woglom. New Haven: Yale University Press.

Mac Lane, S. 1991 (reprint). *Proof, Truth, and Confusion*. The 1982 Nora and Edward Ryerson Lecture at The University of Chicago. Reprinted in *Solstice: An Electronic Journal of Geography and Mathematics*, Volume II, Number 2. http://www-personal.umich.edu/%7Ecopyrght/image/monog15/fulltext.pdf

Mac Lane, S. and G. Birkhoff. 1967. *Algebra*. New York: Macmillan.

MacEachren, A. M. 1995. *How Maps Work*. New York: The Guilford Press.

MacEachren, A. M. 1994. *Some Truth with Maps: A Primer on Symbolization & Design*. University Park: The Pennsylvania State University.

Maher, M. 2009. The Buffer Wizard in ArcMap. http://blogs.esri.com/esri/arcgis/2009/07/15/the-buffer-wizard-in-arcmap/

Mandelbrot, B. 1983. *The Fractal Geometry of Nature*. San Francisco: W. H. Freeman.

MapMaker song: http://wiki.openstreetmap.org/wiki/Mapmaker_Song

Math Is Fun: Accuracy and Precision. http://www.mathsisfun.com/accuracy-precision.html

Mayer, H. M. and R. C. Wade. 1973. *Chicago: Growth of a Metropolis*. Chicago: University of Chicago Press.

Michigan Department of Technology, Management & Budget, http://www.michigan.gov/cgi/0,4548,7-158-52927_53037_12540_13084—,00.html. School Districts and Intermediate School Districts Boundary Map.

Minkowski, H. (1907/1915). Das Relativitätsprinzip. *Annalen der Physik* 352 (15): 927–938.

Moellering, H. and R. L. Hogan, Eds. 1997. *Spatial Database Transfer Standards. Volume 2.* London: Elsevier Science Ltd.

Monmonier, M. 1993. *Mapping It Out.* Chicago: University of Chicago Press.

Monmonier, M. and deBlij, H. 1996. *How to Lie with Maps.* Second Edition. Chicago: University of Chicago Press.

Muehrcke, P. C. and J. O. Muehrcke. 1992. Map Use: Reading, Analysis, Interpretation. Madison, Wisconsin: JP Publications: 210–216.

Murray, A. and J. Gottsegen, Jonathan. The Influence of Data Aggregation on the Stability of Location Model Solutions. http://www.ncgia.ucsb.edu/~jgotts/murray/murray.html

National Atlas. What is a Map Projection. http://www.nationalatlas.gov/articles/mapping/a_projections.html

National Map Accuracy Standards. US Geological Survey. (http://egsc.usgs.gov/isb/pubs/factsheets/fs17199.html)

Newman, J. R. 1956. *The World of Mathematics.* (Four-volume set.) New York: Simon & Schuster.

Newton, I. 1687. Law of universal gravitation. *Philosophiae Naturalis Principia Mathematica.*

Night map of the whole Earth. http://www.nightearth.com/

Nyerges, T., H. Couclelis, and R. McMaster. 2011. The SAGE Handbook of GIS and Society. London: SAGE Publications, Ltd. http://www.uk.sagepub.com/books/Book231755

Nystuen, J. D. 1966. *Effects of boundary shape and the concept of local convexity.* Michigan Interuniversity Community of Mathematical Geographers (unpublished). Reprinted, Ann Arbor: Institute of Mathematical Geography. http://www-personal.umich.edu/~copyrght/image/micmog.html

Nystuen, J. D. and D. Brown. 2001. Coordinators, UM GIS Lecture Series. Lecture by Arthur Getis: "Spatial Analytic Approaches to the Study of the Transmission of Disease: Recent Results on Dengue Fever in Iquitos, Peru."

Okabe, A., Boots, B., and Sugihara, K. 1992. *Spatial Tessellations: Concepts and Applications of Voronoi Diagrams.* Wiley Series in Probability and Mathematical Statistics. Chichester: John Wiley & Sons.

Olmstead, A. T. 1948. *History of the Persian Empire.* Chicago: University of Chicago Press. Found in the following link, accessed on July 18, 2012. http://www.hsc.csu.edu.au/ancient_history/societies/near_east/persian_soc/persiansociety.html

Olson, J. M. 1975. Experience and the improvement of cartographic communication. *Cartographic Journal* 12 (2): 94–108.

Ord, J. K. and A. Getis. 2001. Testing for Local Spatial Autocorrelation in the Presence of Global Autocorrelation. *Journal of Regional Science*, Vol. 41, No. 3, pp. 411–432.

Ovenden, M. 2007. Transit Maps of the World. New York: Penguin Books.

Paret, M. 2012. Using the Mean in Data Analysis: It's not always a Slam Dunk. http://blog.minitab.com/blog/michelle-paret/using-the-mean-its-not-always-a-slam-dunk

Pearson, F. 1990. *Map Projections: Theory and Applications.* Boca Raton, FL: CRC Press.

Perkal, J. 1966. *An attempt at objective generalization.* Michigan Interuniversity Community of Mathematical Geographers (unpublished). Reprinted, Ann Arbor: Institute of Mathematical Geography. http://www-personal.umich.edu/~copyrght/image/micmog.html

Peterson, M. P. 1995. *Interactive and Animated Cartography.* New Jersey: Prentice Hall.

Philbrick, Allen K. *This Human World.* Reprint. Ann Arbor: Institute of Mathematical Geography, 1986. Persistent URL (URI): http://hdl.handle.net/2027.42/58258

Puu, T. 1997. *Mathematical Location and Land Use Theory.* New York: Springer-Verlag.

Quine, W. V. O. 1969. *Ontological Relativity, and Other Essays.* New York: Columbia University Press.

Rayle, R. 2012. Tornado Track Presentation: http://www.mylovedone.com/GooglEarthDay2012/

Reilly, W. J. 1929. Methods for the Study of Retail Relationships. *University of Texas, Bulletin* No 2944.

Reilly, W. J. 1931. The Law of Retail Gravitation. New York: W. J. Reilly, Inc.

Rhynsburger, D. 1973. Analytic delineation of Thiessen polygons. *Geographical Analysis*, 5, 133–144.

Robelen, E. 2012. Draft Standards Neglect Computer Science, Coalition Says. http://blogs. edweek.org/edweek/curriculum/2012/06/the_first_public_draft_of.html?intc=es

Robinson, A. H. 1982. *Early Thematic Mapping: In the History of Cartography*. Chicago: The University of Chicago Press.

Robinson, A. H. 1960. *Elements of Cartography*, 2nd Edition. New York: John Wiley & Sons.

Robinson, A. H. 1953. *Elements of Cartography*. New York: John Wiley & Sons.

Robinson, A. H., J. L. Morrison, P. C. Muehrcke, A. J. Kimerling, and S. C. Guptill. 1995. *Elements of Cartography*. Sixth edition. New York: John Wiley & Sons.

Sammataro, D. with S. L. Arlinghaus. 2011. Varroa Mite Project. *Solstice: An Electronic Journal of Geography and Mathematics*. Ann Arbor: Institute of Mathematical Geography. http://www.mylovedone.com/image/solstice/win11/VarroaGlobal.kmz

Sammataro, D. and S. Arlinghaus. 2010. The Quest to Save Honey: Tracking Bee Pests Using Mobile Technology. *Solstice: An Electronic Journal of Geography and Mathematics*. Volume XXI, No. 2. http://www.mylovedone.com/image/solstice/win10/SammataroandArlinghaus.html

Sammataro, D. with S. L. Arlinghaus. 2009. Bee Ranges and Almond Orchard Locations: Contemporary Visualization. *Solstice: An Electronic Journal of Geography and Mathematics*. Ann Arbor: Institute of Mathematical Geography. http://www-personal. umich.edu/%7Ecopyrght/image/solstice/sum09/BeeRangesAlmonds.html

Sammataro, D. 1998. in S. L. Arlinghaus, W. D. Drake, J. D. Nystuen, with data input from A. Laug, K. S. Oswalt, and D. Sammataro. Animaps. *Solstice: An Electronic Journal of Geography and Mathematics*. Volume IX, Number 1. http://www-personal.umich. edu/%7Ecopyrght/image/solstice/sum98/animaps.html

Science Daily: June 13, 2012. Toddler Spatial Knowledge Boosts Understanding of Numbers. http://www.sciencedaily.com/releases/2012/06/120613102005.htm

Slocum, T. 2003. *Thematic Cartography and Geographic Visualization*. New Jersey: Prentice Hall.

Snyder, J. P. 1987. *Map Projections—A Working Manual*. US Geological Survey Professional Paper 1395. Washington, DC: US Government Printing Office.

Snyder, J. P. 1993. *Flattening the Earth: Two Thousand Years of Map Projections*. Chicago: University of Chicago Press.

Snyder, J. P. and P. M. Voxland. 1994. An Album of Map Projections. US Geological Survey Professional Paper 1453. Washington, DC: US Government Printing Office.

Stewart, J. Q. 1947. Empirical Mathematical Rules Concerning the Distribution and Equilibrium of Population. *Geographical Review*, Volume 37, 461–486.

Thiessen, A. H. and J. C. Alter. 1911. Climatological Data for July, 1911: District No. 10, Great Basin. *Monthly Weather Review,* July: 1082–1089.

Thompson, D'A. W. 1917. *On Growth and Form*. Abridged Edition, edited by J. T. Bonner. Cambridge at the University Press, 1961, first published, 1917.

Thrall, G. I. Series Editor. *Classics in Regional Science*. Scientific Geography, Sage Publications. Regional Research Institute. http://rri.wvu.edu/web_book/booklist/classics-in-regional-science-2

TIGER page, US Census Bureau http://www.census.gov/geo/www/tiger/shp.html

Tissot, N. A. 1881. *Mémoire sur la représentation des surfaces et les projections des cartes géographiques*. Paris: Gauthier-Villars.

Tobler, W. R. 2001. Spherical Measures without Spherical Trigonometry. *Solstice: An Electronic Journal of Geography and Mathematics*. Volume XII, No. 2. Ann Arbor: Institute of Mathematical Geography.

Tobler, W. 1987. Measuring Spatial Resolution, *Proceedings, Land Resources Information Systems Conference*, Beijing, pp. 12–16.

Tobler, W. 1988. Resolution, Resampling, and All That, pp. 129–137 of H. Mounsey and R. Tomlinson, eds., *Building Data Bases for Global Science*, London: Taylor & Francis.

Tobler, W. R. 1973. Tobler Hyperelliptical Projection. http://www.csiss.org/map-projections/Pseudocylindrical/Hyperelliptical.pdf

Tobler, W. R. 1970. A computer movie simulating urban growth in the Detroit region. *Economic Geography*, 46 (2): 234–240. http://en.wikipedia.org/wiki/Tobler%27s_first_law_ of_geography

Tobler, W. R. 1961. World Map on a Moebius Strip, *Surveying and Mapping*, XXI, 4, 486.

Tobler, W. R. 1961. Map Transformations of Geographic Space, PhD Thesis, University of Washington, Seattle. University Microfilms No. 61 – 4011.

Tufte, E. R. 1990. *Envisioning Information*. Cheshire, Connecticut: Graphics Press.

Unwin, D. J. 2010. Numbers aren't nasty: A workbook of spatial concepts. http://www.teachspatial.org/sites/teachspatial.org/files/Unwin_WorkbookOfSpatialConcepts.pdf

US Census Bureau. Census Data Access Tools. May need Microsoft Silverlight installed.

US Department of the Interior. Mapmaker. National Atlas. http://nationalatlas.gov

USGS Map Projections Poster. http://egsc.usgs.gov/isb/pubs/MapProjections/projections.html

USGS 2012. The National Map. US Topo.

Varenius, B. 1650. *Geographia Generalis*.

Venn, J. 1880. On the Diagrammatic and Mechanical Representation of Propositions and Reasonings. *Philosophical Magazine and Journal of Science*. 5, 10, (59).

von Thünen, J. H. 1966. *The Isolated State*, trans. Carla M. Wartenberg, edited with an Introduction by Peter Hall, London: Pergamon Press.

Vries, H. de. 1906. *Species and Varieties: Their Origin by Mutation*. Second edition. Chicago: The Open Court Publishing Company Ed. by D. T. MacDougal.

Warntz, W. 1965. *A note on surfaces and paths and applications to geographical problems*. Number 6 in the Papers of the Michigan Inter-university of Mathematical Geographers. http://www-personal.umich.edu/~copyrght/image/micmog.html. Housed in permanent archive, DeepBlue: http://deepblue.lib.umich.edu/handle/2027.42/58219

Washtenaw County, City of Ann Arbor GIS. Map Washtenaw (interactive). May need Microsoft Silverlight installed.

Washtenaw County website commentary on preparedness. http://www.ewashtenaw.org/government/departments/emergency_management

Weber, A. 1928. *Theory of the Location of Industries*, trans. C. J. Friedrich. Chicago: The University of Chicago Press.

Weissstein, E. W. 1999. Sinusoidal Projection. From *MathWorld*—A Wolfram Web Resource. http://mathworld.wolfram.com/SinusoidalProjection.html

Weissstein, E. W. 1999. Mollweide Projection. From *MathWorld*—A Wolfram Web Resource. http://mathworld.wolfram.com/MollweideProjection.html

Weissstein, E. W. 1999. Mercator Projection. From *MathWorld*—A Wolfram Web Resource. http://mathworld.wolfram.com/MercatorProjection.html

Weissstein, E. W. 1999. Conic Projection. From *MathWorld*—A Wolfram Web Resource. http://mathworld.wolfram.com/ConicProjection.html

Wood, J. 2000. GPS Maps. http://www.gpsdrawing.com/maps.html. Samples on EveryTrail: http://blog.everytrail.com/?p=61

Yang, Q. H., J. P. Snyder, and W. R. Tobler. 2000. *Map Projection Transformation: Principles and Applications*. London: Taylor & Francis.

Zipf, G. K. 1946. The P1 P2/D Hypothesis: On the Intercity Movement of Persons. *American Sociological Review*. Volume 11.

Zipf, G. K. 1949. *Human Behavior and the Principle of Least Effort*. New York: Hafner.

Index

horizontal, 18
North American, 18
World Geodetic System of 1984, 18
DD, *see* Decimal degrees (DD)
Decimal degrees (DD), 6, 37, 249
    latitude and longitude, 13
Decimal minutes (DM), 43
Degree (°), 6
Degree Confluence Project, 209–210, 249;
    see also Map projections
Degrees, minutes, and seconds (DMS), 41
Demographic analysis, 167, 249
    using ArcGIS desktop for, 170
Demographic gravitation, 224
Diametral planes, 2; *see also* Sphere
    Cartesian grid, 6
Digital Line Graphs (DLG), 141
Digital mapping, 71; *see also* Mapping;
    Raster mapping
    issues, 73
    raster image, 71, 72
    vector image, 71, 72
Digital maps, 141
    crossings, 139
    GIS-based, 13
Digitizing, 249
Directional trend, 249
Dispersion, 249
Distributive law, 98–99
DLG, *see* Digital Line Graphs (DLG)
DM, *see* Decimal minutes (DM)
DMS, *see* Degrees, minutes, and seconds
    (DMS)
Dot density map, 115, 249; *see also* Scale
    absolute representation, 117
    clustering, 115–117
    construction of, 115–117
    dot scatter, 115, 116
    optimization principle, 119
    practice exercises, 123–126
    projection principle, 119–120
    randomizing principle, 118–119
    relative representation, 117
    scale principle, 119
    theory, 118
Duality, principle of, 228

E

Earth, 17; *see also* Google Earth
    annual revolution of, 11
    physical, 17
    reference system for, 5
    sunlight intensity and poles, 12
    tilt of Earth's polar axis, 11

Earth circumference; *see also* Equinoxes
    angle measurement, 36
    assumptions made, 35
    by Eratosthenes, 34
    practice exercises, 37–41
Earth coordinate systems, 2; *see also*
    Geometry of sphere
    absolute system, 6
    antipodal points, 3
    Cartesian coordinate system, 9–10
    diametral planes, 2
    Earth-sphere graticule, 7
    geodesics, 2
    great circle, 2, 3
    location of *P*, 5
    reference lines, 3, 4
    small circle, 2, 3
    standard circular measure, 6
Earth models, 17, 18
    datum, 17–19
    in geography and mathematics, 17
    geoid, 17
    physical Earth, 17
    reference ellipsoid, 17
Earth-sphere; *see also* Coordinate systems;
    Earth circumference
    common coordinate systems, 43–48
    graticule, 7
    location and measurement, 33
    position measurement on Earth surface,
        41–43
    practice exercises, 37–41, 47–48, 50–51,
        51–52, 52–54
    related theory and practice, 54–55
    relative and absolute location, 32–33
    visual trigonometry, 48
Earth's seasons, 10; *see also* Equinoxes
    Earth annual revolution, 11
    primary cause of, 10
    sunlight intensity and poles, 12
    tilt of Earth's polar axis, 11
Eastings, 44
Eckert IV projection, 212
Ecliptic plane, 10
Elevation, 249
Elliptic geometry, 215
Epicenters, preliminary determinations
    of, 14
Equal area projection, 119, 249
Equal interval classification method, 134;
    see also Data partitioning
    merits and limitations of, 133
    practice exercises, 145–150
Equator, 199, 206
Equidistant, 249

USGS, *see* United States Geological Survey (USGS)
UTM, *see* Universal Transverse Mercator (UTM)

V

Varroa mite, 144
VBA, *see* Visual Basic for Applications (VBA)
Vector format, 73, 253; *see also* Raster format
Vector image, 71, 72; *see also* Raster image
    limitations, 73
    merits, 72
    vector data resolution, 74
    Venn diagram, 69
Verbal Scale, 112; *see also* Scale
Visual Basic for Applications (VBA), 173
Visual trigonometry, 48
Voxel, 93, 251, 253; *see also* Pixel
    color voxel space, 94, 95

W

WAAS, *see* Wide Area Augmentation System (WAAS)

Warntz map, 224
Web mapping, 253; *see also* Geographic Information System (GIS)
    TIGERweb mapping service, 120
Web Mercator Auxiliary Sphere, 253
    projection, 208–209; *see also* Map projections
Web-based GIS, 101, 253; *see also* ArcGIS Online
    for path analysis, 20
Weighted mean center, 188, 253
WGS, *see* World Geodetic System (WGS)
Wide Area Augmentation System (WAAS), 38
World from space, the, 217; *see also* Google Earth; Map projections
World Geodetic System (WGS), 82
    in GPS receivers, 18

Z

Zero point, 17
Zones of proximity, 66; *see also* Euclidean buffers; Geographic Information System (GIS)